MATHEMATICS, MODELS, AND MODALITY

John Burgess is the author of a rich and creative body of work which seeks to defend classical logic and mathematics through counter-criticism of their nominalist, intuitionist, relevantist, and other critics. This selection of his essays, which spans twenty-five years, addresses key topics including nominalism, neo-logicism, intuitionism, modal logic, analyticity, and translation. An introduction sets the essays in context and offers a retrospective appraisal of their aims. The volume will be of interest to a wide range of readers across philosophy of mathematics, logic, and philosophy of language.

JOHN P. BURGESS is Professor in the Department of Philosophy, Princeton University. He is co-author of *A Subject With No Object* with Gideon Rosen (1997) and *Computability and Logic*, 5th edn with George S. Boolos and Richard C. Jeffrey (2007), and author of *Fixing Frege* (2005).

MATHEMATICS, MODELS, AND MODALITY

Selected Philosophical Essays

JOHN P. BURGESS

CAMBRIDGE UNIVERSITY PRESS
Cambridge, New York, Melbourne, Madrid, Cape Town,
Singapore, São Paulo, Delhi, Tokyo, Mexico City

Cambridge University Press
The Edinburgh Building, Cambridge CB2 8RU, UK

Published in the United States of America by Cambridge University Press, New York

www.cambridge.org
Information on this title: www.cambridge.org/9780521189675

First published 2008
First paperback edition 2011

A catalogue record for this publication is available from the British Library

ISBN 978-0-521-88034-3 Hardback
ISBN 978-0-521-18967-5 Paperback

Dedicated to the memory of my sister
Barbara Kathryn Burgess

Contents

Preface

The present volume contains a selection of my published philosophical papers, plus two items that have not previously appeared in print. Excluded are technical articles, co-authored works, juvenilia, items superseded by my published books, purely expository material, and reviews. (An annotated partial bibliography at the end of the volume briefly indicates the contents of such of my omitted technical papers as it seemed to me might interest some readers.) The collection has been divided into two parts, with papers on philosophy of mathematics in the first, and on other topics in the second; references in the individual papers have been combined in a single list at the end of the volume. Bibliographic data for the original publication of each item reproduced here are given source notes on pp. xi–xiii, to which the notes of personal acknowledgment, dedications, and epigraphs that accompanied some items in their original form have been transferred; abstracts that accompanied some items have been omitted.

It has become customary in volumes of this kind for the author to provide an introduction, relating the various items included to each other, as an editor would in an anthology of contributions by different writers. I have fallen in with this custom. The remarks on the individual papers in the introduction are offered primarily in the hope that they may help direct readers with varying interests to the various papers in the collection that should interest them most. But such introductions also serve another purpose: they provide an opportunity for an author to note any changes of view since the original publication of the various items, thus reducing any temptation to tamper with the text of the papers themselves on reprinting. I have made only partial use of the opportunity to note changes in view, but nonetheless I have felt no temptation to make substantial changes in the papers, since my own occasional historical research has convinced me of the badness of the practice of revising papers on reprinting.

I have tried to acknowledge in each individual piece those to whom I have been most indebted in connection with that item, though I am sure there are some I have unintentionally neglected, whose pardon I must beg. Here I would like to acknowledge those who have been helpful specifically with the preparation of the present collection: Hilary Gaskin, who first suggested such a volume, and Joanna Breeze, along with Gillian Dadd and the rest of the staff who saw the work through publication.

Source notes

"Numbers and ideas" was first delivered orally as part of a public debate at the University of Richmond (Virginia), 1999. Ruben Hersh argued *for* the thesis "Resolved: that mathematical entities and objects exist within the world of shared human thoughts and concepts." I argued against. It was first published in a journal for undergraduates edited at the University of Richmond (England), the *Richmond Journal of Philosophy*, volume 1 (2003), pp. 12–17. (There is no institutional connection between the universities of the two Richmonds, and my involvement with both is sheer coincidence.)

"Why I am not a nominalist" was first delivered orally under the title "The nominalist's dilemma," to the Logic Club, Catholic University of Nijmegen, 1981. It was first published in the *Notre Dame Journal of Formal Logic*, volume 24 (1983), pp. 93–105.

"Mathematics and *Bleak House*" was first delivered orally at a symposium "Realism and anti-realism" at the Association for Symbolic Logic meeting, University of California at San Diego, 1999. The other symposiast was my former student Penelope Maddy, and the Dickensian title of my paper is intended to recall the Dickensian title of her earlier review, "Mathematics and *Oliver Twist*" (Maddy 1990). First published in *Philosophia Mathematica*, volume 12 (2004), pp. 18–36.

"Quine, analyticity, and philosophy of mathematics" was first delivered orally at the conference "Does Mathematics Require a Foundation?," Arché Institute, University of St. Andrews, 2002. Identified in its text as a sequel to the preceding item, this paper circulated in pre-publication draft under the title "Mathematics and *Bleak House*, II." First published in the *Philosophical Quarterly*, volume 54 (2004), pp. 38–55.

"Being explained away" is a shortened version (omitting digressions on technical matters) of a paper delivered orally to the Department of Philosophy, University of Southern California, 2004. (I wish not only to thank that department for the invitation to speak, but especially to thank

Stephen Finlay, Jeff King, Zlatan Damnjanovic, and above all Scott Soames for their comments and questions, as well as for their hospitality during my visit.) It was first published in the *Harvard Review of Philosophy*, volume 13 (2005), pp. 41–56.

"*E pluribus unum*" evolved from a paper "From Frege to Friedman" delivered orally at the Logic Colloquium of the University of Pennsylvania and the Department of Logic and Philosophy of Science at the University of California at Irvine. It was first published in *Philosophia Mathematica*, volume 12 (2004), pp. 193–221. (I am grateful to Harvey Friedman for introducing me to his recent work on reflection principles, to Kai Wehmeier and Sol Feferman for drawing my attention to the earlier work of Bernays on that topic, and to Penelope Maddy for pressing the question of the proper model theory for plural logic, which led me back to the writings of George Boolos on this issue. From Feferman I also received valuable comments leading to what I hope is an improved exposition.)

"Logicism: a new look" was first delivered orally at the conference marking the inauguration of the UCLA Logic Center, and later (under a different title) as part of the annual lecture series of the Center for Philosophy of Science, University of Pittsburgh, both in 2003. It has not previously been published.

"Tarski's tort" was first delivered orally at Timothy Bays' seminar on truth, Notre Dame University, Saint Patrick's Day, 2005. It was previously unpublished. The paper should be understood as dedicated to my teacher Arnold E. Ross, mentioned in its opening paragraphs.

"Which modal logic is the right one?" was first delivered orally at the George Boolos Memorial Conference, University of Notre Dame, 1998. It was first published in the *Notre Dame Journal of Formal Logic*, volume 40 (1999), pp. 81–93, as part of a special issue devoted to the proceedings of that conference. Like all the conference papers, mine was dedicated to the memory of George Boolos.

"Can truth out?" was first delivered orally under the title "Fitch's paradox of knowability" as a keynote talk at the annual Princeton–Rutgers Graduate Student Conference in Philosophy, 2003. It was first published in Joseph Salerno, ed., *New Essays on Knowability*, Oxford: Oxford University Press (2007). The paper originally bore the epigraph "Truth will come to light; murder cannot be hid long; a man's son may, but at the length truth will out" (*Merchant of Venice* II: 2). Thanks are due to Michael Fara, Helge Rückert, and Timothy Williamson for perceptive comments on earlier drafts of this note.

"*Quinus ab omni naevo vindicatus*" was first delivered orally to the Department of Philosophy, MIT, 1997. It was first published in Ali Kazmi, ed., *Meaning and Reference*, *Canadian Journal of Philosophy Supplement*, volume 23 (1998), pp. 25–65. (The present paper is a completely rewritten version of an unpublished paper, "The varied sorrows of modality, part II." I am indebted to several colleagues for information used in writing that paper, and for advice given on it once written, and I would like to thank them all – Gil Harman, Dick Jeffrey, David Lewis – even if the portions of the paper with which some of them were most helpful have disappeared from the final version. But I would especially like to thank Scott Soames, who was most helpful with the portions that have *not* disappeared.)

"Translating names" was first published in *Analysis*, volume 65 (2005), pp. 96–204. I am grateful to Pierre Bouchard and Paul Égré for linguistic information and advice.

"Relevance: a fallacy?" was first published in the *Notre Dame Journal of Formal Logic*, volume 22 (1981), pp. 76–84. Its sequels were Burgess (1983c) and Burgess (1984b).

"Dummett's case for intuitionism" was first published in *History and Philosophy of Logic*, volume 5 (1984), pp. 177–194. The paper originally bore the epigraph from Chairman Mao "Combat Revisionism!" I am indebted to several colleagues and students for comments, and especially to Gil Harman, who made an earlier draft of this paper the topic for discussion at one session of his summer seminar. Comments by editors and referees led to what it is hoped are clearer formulations of many points.

Introduction

ABOUT "REALISM"

A word on terminology may be useful at the outset, since it is pertinent to many of the papers in this collection, beginning with the very first. The label "realism" is used in two very different ways in two very different debates in contemporary philosophy of mathematics. For nominalists, "realism" means acceptance that there exist entities, for instance natural or rational or real numbers, that lack spatiotemporal location and do not causally interact with us. For neo-intuitionists, "realism" means acceptance that statements such as the twin primes conjecture may be true independently of any human ability to verify them. For the former the question of "realism" is ontological, for the latter it is semantico-epistemological. Since the concerns of nominalists and of neo-intuitionists are orthogonal, the double usage of "realism" affords ample opportunity for confusion.

The arch-nominalists Charles Chihara and Hartry Field, for instance, are anti-intuitionists and "realists" in the neo-intuitionists' sense. They do not believe there are any unverifiable truths about numbers, since they do not believe there are any numbers for unverifiable truths to be about. But they do believe that the facts about the possible production of linguistic expressions, or about proportionalities among physical quantities, which in their reconstructions replace facts about numbers, can obtain independently of any ability of ours to verify that they do so. Michael Dummett, the founder of neo-intuitionism, was an early and forceful anti-nominalist, and though he calls his position "anti-realism," he and his followers are "realists" in the nominalists' sense, accepting some though not all classical existence theorems, namely those that have constructive proofs, and agreeing that it is a category mistake to apply spatiotemporal or causal predicates to mathematical subjects.

On top of all this, even among those of us who are "realists" in both senses there are important differences. *Metaphysical* realists suppose, like

Galileo and Kepler and Descartes and other seventeenth-century worthies, that it is possible to get behind all human representations to a God's-eye view of ultimate reality as it is in itself. When *they* affirm that mathematical objects transcending space and time and causality exist, and mathematical truths transcending human verification obtain, they are affirming that such objects exist and such truths obtain *as part of ultimate metaphysical reality* (whatever that means). *Naturalist* realists, by contrast, affirm only (what even some self-described anti-realists concede) that the existence of such objects and obtaining of such truths is an implication or presupposition of science and scientifically informed common sense, while denying that philosophy has any access to exterior, ulterior, and superior sources of knowledge from which to "correct" science and scientifically informed common sense. The naturalized philosopher, in contrast to the alienated philosopher, is one who takes a stand as a citizen of the scientific community, and not a foreigner to it, and hence is prepared to reaffirm while doing philosophy whatever was affirmed while doing science, and to acknowledge its evident implications and presuppositions; but only the metaphysical philosopher takes the status of what is affirmed while doing philosophy to be a revelation of an ultimate metaphysical reality, rather than a human representation that is the way it is in part because a reality outside us is the way it is, and in part because we are the way we are.

My preferred label for my own position would now be "naturalism," but in the papers in this collection, beginning with the first, "realism" often appears. Were I rewriting, I might erase the R-word wherever it occurs; but as I said in the preface above, I do not believe in rewriting when reprinting, so while in date of composition the papers reproduced here span more than twenty years, still I have left even the oldest, apart from the correction of typographical errors, just as I wrote them. *Quod scripsi, scripsi.*

This collection begins with five items each pertinent in one way or another to nominalism and the problem of the existence of abstract entities. The term "realism" is used in an ontological sense in the first of these, "Numbers and ideas" (2003). This paper is a curtain-raiser, a lighter piece responding to certain professional mathematicians turned amateur philosophers who propose a cheap and easy solutions to the problem. According to their proposed compromise, numbers exist, but only "in the world of ideas." Since acceptance of this position would render most of the professional literature on the topic irrelevant, and since the amateurs often offer unflattering accounts of what they imagine to be the reasons why professionals do not accept their simple proposal, I thought it worthwhile to accept an invitation to try to state, for a general audience, our real reasons, which go back to Frege. The distinction insisted upon in this paper, between the kind of thing it makes sense to say about a number and the kind of thing it makes sense to say about a mental representation of a number (and the distinction, which exactly parallels that between the two senses of "history," between mathematics, the science, and mathematics, its subject matter) is presupposed throughout the papers to follow.

Some may wonder where my emphatic rejection of "idealism or conceptualism" in this paper leaves intuitionism. The short answer is that I leave intuitionism entirely out of account: I am concerned in this paper with descriptions of the mathematics we have, not prescriptions to replace it with something else. Intuitionism is orthogonal to nominalism, as I have said, and issues about it are set aside in the first part of this collection. I will add that, though I do not address the matter in the works reprinted here, my opinion is that Frege's anti-psychologistic and anti-mentalistic points raise some serious difficulties for Brouwer's original version of intuitionism, but no difficulties at all for Dummett's revised version. Neither opinion should be controversial. Dummett's producing a version immune to Fregean criticism can hardly surprise, given that the founder of neo-intuitionism is also the dean of contemporary Frege studies. That Brouwer's version, by contrast, faces serious problems was conceded even by so loyal a disciple as Heyting, and all the more so by contemporary neo-intuitionists.

AGAINST HERMENEUTIC AND REVOLUTIONARY NOMINALISM

"Why I am not a nominalist" (1983) represents my first attempt to articulate a certain complaint about nominalists, namely, their unclarity about the distinction between *is* and *ought*. It was this paper that first introduced a distinction between *hermeneutic* and *revolutionary* nominalism. The formulations a decade and a half later in *A Subject With No Object* (Burgess and Rosen, 1997) are, largely owing to my co-author Gideon Rosen, who among other things elaborated and refined the hermeneutic/revolutionary distinction, more careful on many points than those in this early paper. This piece, however, seemed to me to have the advantage of providing a more concise, if less precise, expression of key thoughts underlying that later book than can be found in any one place in the book itself. Inevitably I have over the years not merely elaborated but also revised (often under Rosen's influence) some of the views expressed in this early article.

First, the brief sketches of projects of Charles Chihara and Hartry Field in the appendix to the paper (which I include on the recommendation of an anonymous referee, having initially proposed dropping it in the reprinting) are in my present opinion more accurate as descriptions of aspirations than of achievements, and even then as descriptions only to a first approximation; moreover the later approach of Geoffrey Hellman is not discussed at all. My ultimate view of the technical side of the issue is given in full detail in the middle portions of *A Subject*, superseding several earlier technical papers.

Further, though I still see no serious linguistic evidence in favor of any hermeneutic nominalist conjectures, I no longer see the absence of such evidence as the main objection to them. For reasons that in essence go back to William Alston, such conjectures lack relevance *even if correct*. Even if we grant that "There are prime numbers greater than a million" does just mean, say, "There could have existed prime numerals greater than a million," the conclusion that should be drawn is that "Numbers exist" means "Numerals could have existed," and is therefore *true*, as anti-nominalists have always maintained, and not *false*, as nominalists have claimed. There is no threat at all to a naturalist version of anti-nominalism in such translations, though there might be to a metaphysical version. This line I first developed in a very belatedly published paper (Burgess 2002a) of which a condensed version was incorporated into *A Subject*.

Finally, I now recognize that there is a good deal more to be said for the position I labeled "instrumentalism" than I or almost anyone active in the field was prepared to grant back in the early 1980s when I wrote "Why I am not," or even in the middle 1990s, when I wrote my contributions to *A Subject*. The position in question is that of those philosophers who speak with the vulgar in everyday and scientific contexts, only to deny on entering the philosophy room that they meant what they said seriously. This view is now commonly labeled "fictionalism," and it deserves more discussion than it gets in either "Why I am not" or *A Subject*. It should be noted that while I originally opposed fictionalism (or instrumentalism) to both the revolutionary and hermeneutic positions, Rosen has correctly pointed out that fictionalism itself comes in a revolutionary version (this is the attitude philosophers *ought to* adopt) and a hermeneutic version (this is the attitude commonsense and scientific thinkers *already do* adopt). What I originally called the "hermeneutic" position should be called the "content-hermeneutic" position, and the hermeneutic version of fictionalism the "attitude-hermeneutic" position, in Rosen's refined terminology.

On two points my view has not changed at all over the past years. First, while nominalists would wish to blur what for Rosen and myself is a key distinction, and avoid taking a stand on whether they are giving a description of the mathematics we already have (hermeneutic) or a prescription for a new mathematics to replace it (revolutionary), gesturing towards a notion of "rational reconstruction" that would somehow manage to be neither the one nor the other, I did not think this notion had been adequately articulated when I first took up the issue of nominalism, and I have not found it adequately articulated in nominalist literature of the succeeding decades.

Second, as to the popular epistemological arguments to the effect that even if numbers or other objects "causally isolated" from us do exist, we cannot *know* that they do, I have not altered the opinions that I expressed in my papers Burgess (1989) and the belatedly published Burgess (1998b), and that Rosen expressed in his dissertation, and that the two of us jointly expressed in *A Subject*. The epistemological argument, according to which belief in abstract objects, even if conceded to be implicit in scientific and commonsense thought, and even if perhaps true – for the aim of going epistemological is precisely to avoid direct confrontation over the question of the *truth* of anti-nominalist existence claims – cannot constitute *knowledge*, surely is not intended as a Gettierological observation about the gap between justified true belief and what may properly be called *knowledge*. It follows that it must be an issue about *justification*; and here to the naturalized anti-nominalist the nominalist appears simply to be substituting some extra-, supra-, praeter-scientific *philosophical* standard of justification for the ordinary standards of justification employed by science and common sense: the naturalist anti-nominalist's answer to nominalist skepticism about mathematics is skepticism about philosophy's supposed access to such non-, un-, and anti-scientific standards of justification.

AGAINST FICTIONALIST NOMINALISM

Returning to the issue of fictionalism, in our subsequent work Rosen and I have generally dealt with it separately and in our own ways. A chapter bearing the names of Rosen and myself, "Nominalism reconsidered," does appear in Stuart Shapiro's *Handbook of Philosophy of Mathematics and Logic* (2005), and it is a sequel to our book adding coverage of fictionalist nominalism, with special reference to the version vigorously advocated over the past several years by Steve Yablo; but this chapter is substantially Rosen's work, my contributions being mainly editorial.

My own efforts to address a fictionalist position are to be found rather in "Mathematics and *Bleak House*," which revisits, in a sympathetic spirit, Rudolf Carnap's ideas on the status of ontological questions and nominalist theses. Neo-Carnapianism is on the rise, and I am happy to be associated with it, though like any other neo-Carnapian I have my differences with my fellow neo-Carnapians. "Quine, analyticity, and philosophy of mathematics" can be read as a sequel to the *Bleak House* paper (it was written much later, though owing to various accidents both came out in the same year, 2004). It revisits the famous exchange between Carnap and Quine on ontology, again in a spirit sympathetic to Carnap.

Carnap thought there was a separation to be made between analytic questions about what is the content of a concept such as that of number, and pragmatic questions about why we accept such a concept for use in scientific theorizing and commonsense thought. Quine denied there was in theory any sharp separation to be made. I argue that there is in practice at least a fuzzy one. I also argue that Quine had better acknowledge as much if he is to be able to make any reply to a serious criticism of Charles Parsons. The criticism is that Quine's holist conception of the justification of mathematics – it counts as a branch of science rather than imaginative literature because of its contribution to other sciences – cannot do justice to the *obviousness* of elementary arithmetic.

Though placed in the first half of this volume along with papers about nominalism, the Quine paper can equally well be read more or less independently as a paper in philosophy of language and theory of knowledge about the notion of analyticity, one that just happens to use mathematics and logic as sources of examples. The placement of this paper, and more generally the division of the collection into two parts, should not be taken too seriously.

As any neo-Carnapian will tell you, though Carnap was certainly an anti-nominalist, his position is perhaps better characterized as generally anti-*ontological* rather than specifically anti-*nominalist*. My own general anti-ontologism became finally, fully, and emphatically explicit in "Being explained away" (2005), my farewell to the issue of nominalism. In this retrospective (written for an audience of undergraduate philosophy concentrators) I distinguish what I call scientific *ontics*, a glorified taxonomy of the entities recognized by science, from what I call philosophical *ontosophy*, an impossible attempt to get behind scientific representations to a God's-eye view, and catalogue the metaphysically ultimate furniture of the universe. The error of the nominalists consists, in my opinion, not in ontosophical anti-realism about the abstract, but in ontosophical realism about the concrete – more briefly, the error is simply going in for ontosophy and not resting content with ontics.

In taking leave of the issue of nominalism, I should reiterate the point made briefly at the end of *A Subject*, that from a naturalist point of view there is a great deal to be learned from the projects of Field, Chihara, Hellman, and others. Naturalists, I have said, hold that there is no possibility of separating completely the contributions from the world and the contributions from us in shaping our theories of the world. At most we can get a hint by considering how the theories of creatures like us in a world unlike ours, or the theories of creatures unlike us in a world like ours, might differ from our own theories. The nominalist reconstruals or reconstructions, though implausible when read as hermeneutic, as accounts of the meaning of our theories, and unattractive when read as revolutionary, as rivals competing for our acceptance with those theories, do give a hint of what the theories of creatures unlike us might be like.

Another hint is provided by those monist philosophers who have reconstrued what appear to be predicates applying to various objects as predicates applying to a single subject, the Absolute, with the phrases that seem to refer to the various objects being reconstrued as various adverbial modifiers. Thus "Jack sings and Jill dances" becomes "The Absolute sings jackishly and dances jillishly," while "Someone sings and someone else dances" becomes "The Absolute sings somehow and dances otherhow." What is specifically sketched in "Being explained away" is how this kind of reconstrual can be systematically extended, at least as far as first-order regimentation of discourse can be extended. Of course it is not to be expected that we can fully imagine what it would be like to be an intelligent creature who habitually thought in such alien terms, any more that we can fully imagine what it would be like to be a bat. Nor insofar as we are capable of partially imagining what is not wholly imaginable are formal studies the only aid to imagination. The kind of fiction that stands to metaphysics as science fiction stands to physics – the example I cite in the paper is Borges – may give greater assistance.

FOUNDATIONS OF MATHEMATICS: SET THEORY

As long as mathematicians adhere to the ideal of rigorous proof from explicit axioms, they will face decisions as to which proposed axioms to start from, and which methods of proof to admit. What is conventionally known as "foundations of mathematics" is simply the technical study, using the tools of modern logic, of the effects of different choices. Work in foundations emphatically does not imply commitment to a "foundationalist" philosophical position, or for that matter to *any* philosophical position. In Burgess (1993) I nevertheless argued that work in foundations can be *relevant* to philosophy, and tried to explain how. I will not attempt to summarize the explanation here, except to give this hint: most of the interesting choices of axioms, especially those that are more restrictive rather than the orthodox choice of something like the axioms of Zermelo–Frankel set theory, were originally inspired by positions in the philosophy of mathematics (finitism, constructivism, predicativism, and others). Foundational work helps us appreciate what is at stake in the choice among those restrictive philosophies, and between them and classical orthodoxy.

While the early papers in the first part of this collection are predominantly though not exclusively critical, and the middle papers a mix of critical and positive – I would say "constructive," except that this word has a special meaning in philosophy of mathematics – the last two are, like the bulk of my more technical work, predominantly though not exclusively positive. Though they do not endorse as ultimately correct, they present as deserving of serious and sustained attention three novel approaches to foundations of mathematics, very different in appearance from each other, but not necessarily incompatible.

To the extent that there is an agreed foundation or framework for contemporary pure mathematics, it is provided by something like the Zermelo–Frankel system of axiomatic set theory, in the version including the axiom of choice (ZFC). "*E pluribus unum*" (2004) attempts to combine two insights, one due to Boolos, the other to Paul Bernays, to achieve an improved framework.

The idea taken from Boolos is that plural quantification on the order of "there are some things, the *us*, such that . . ." is a more primitive notion than singular quantification of the type "there is a set or class U of things such that . . ." and that Cantor's transition from the former to the latter was a genuine conceptual innovation, not a mere uncovering of a commitment to set- or class-like entities that had been implicit in ordinary plural talk all along.

Boolos himself had applied this idea to set theory, to suggest, not improved axioms, but an improved formulation of the existing axioms. For there is a well-known awkwardness in the formulation of ZFC, in that two of its most important principles appear not as axioms but as *schemes*, or rules to the effect that all sentences of a certain form are to count as axioms. For instance, *separation* takes the form

$$\forall x \exists y \forall z (z \in y \leftrightarrow z \in x \;\&\; \phi(z))$$

wherein ϕ may be any formula. Needless to say, no one becomes convinced of the correctness of ZFC by becoming convinced separately of each of infinitely many instances of the separation scheme. But the language of ZFC provides no means of formulating the underlying *single, unified* principle. One proposed solution to this difficulty has been to recognize collections of a kind called *classes* that are set-like while somehow failing to be sets. With capital letters ranging over such entities, and with "$z \in U$" written "Uz" to emphasize that the relation of class membership is a kind of belonging that is like set-elementhood and yet somehow fails to be set-elementhood, the separation scheme can be reduce to a single axiom, thus:

$$\forall U \forall x \exists y \forall z(z \in y \leftrightarrow z \in x \;\&\; Uz).$$

But notion of class brings with it difficulties of its own, leaving many hesitant to admit these alleged entities.

The suggestion of Boolos (in my own notation) was to replace singular quantification $\forall U$ or "for any class U of sets ..." over classes by plural quantification $\forall\forall uu$ or "for any sets, the u's ..." and Uz or "z is a member of U" by $z \propto uu$ or "z is one of the u's," thus yielding a formulation in which the only objects quantified over are sets:

$$\forall\forall uu \forall x \exists y \forall z(z \in y \leftrightarrow z \in x \;\&z \propto uu).$$

One may even take a further step and make the notion $x \equiv uu$ or "x is the set of the u's" primitive, with the notion $y \in x$ or "y is an element of x" being defined in terms of it, as $\exists\exists uu(xx \equiv uu \;\&\; y \propto uu)$ or "there are some things that x is the set of, and y is one of them." Such a step was actually taken in a paper by Stephen Pollard (1996) some years before my own, of which I only belated became aware, along with Shapiro (1987) and Rayo and Uzquiano (1999).

The idea taken from Bernays was that an approach incorporating a so-called reflection principle can provide a simpler axiomatization than the standard approach to motivating the axioms of ZFC, and permit the derivation of some further so-called large-cardinal principles that are widely accepted by set theorists, though they go beyond ZFC. The original Bernays approach had the disadvantage of involving "classes" over and above sets, and of requiring a somewhat artificial technical condition in the formulation of the reflection principle. Boolos's plural logic was subject to the objection that, like any version or variant of second-order logic, it lacks a complete axiomatization. I aim to show how the combination of Boolos with Bernays neutralizes these objections.

FOUNDATIONS OF MATHEMATICS: LOGICISM

"Logicism: a new look" (previously unpublished) provides a concise, semi-popular introduction to two alternative approaches to foundations each of which I have examined more fully and technically elsewhere. Each represents a version of the old idea of *logicism*, according to which mathematics is ultimately but a branch of logic. Computational facts such as $2+2=4$, on this view, become abbreviations for logical facts; in this case, the fact that if there exists an F and another and no more, and a G and another and no more, and nothing is both an F and a G, and something is an H if and only if it is either an F or a G, then there exists an H and another and yet another and still yet another, but no more.

One new idea derives from Richard Heck. Frege, the founder of modern logic and modern logicism proposed to develop arithmetic in a grand system of logic of his devising. That system is, in modern notation and to a first approximation, a form of second-order logic, with axioms of *comprehension* and *extensionality*,

$$\exists X \forall x(Xx \leftrightarrow \phi(x))$$
$$\forall X \forall Y(\forall z(Xz \leftrightarrow Yz) \rightarrow (\phi(X) \leftrightarrow \phi(Y)))$$

supplemented by an axiom to the effect that to each second-order entity X there is associated a first-order entity X^* in such a way that we have

$$\forall X \forall Y(X^* = Y^* \leftrightarrow \forall z(Xz \leftrightarrow Yz)).$$

Russell showed that a paradox arises in this system, and also introduced the idea of imposing a restriction of *predicativity* on the comprehension axiom, assuming it only for formulas $\phi(x)$ without bound class variables. Russell proposed a great many other changes, and his overall system diverged greatly from Frege's. Heck was the first to consider closely what would happen if one made *only* the one change just described, and he showed that the resulting system, though weak (and in particular consistent) is strong enough for the minimal arithmetic embodied in the system known in the literature as Q to be developed in it. So a bare minimum of mathematics can be developed on a predicative logicist basis in the manner of Frege. (More technical details as to what can be accomplished along these lines are provided in my book *Fixing Frege* (Burgess 2005b). Since the book and the paper were written there have been important advances by Mihai Ganea and Albert Visser.)

Russell's version of logicism was opposed by Brouwer's intuitionism and also by Hilbert's formalism. The latter was consciously modeled on instrumentalist philosophies of physics, according to which physical theory is a giant instrument for deriving empirical predictions, though theoretical terms and laws in physics in general do not admit direct empirical definitions or meaning. Hilbert's philosophy of mathematics can be represented by a simple proportion:

computational : mathematics :: empirical : physics.

For Hilbert, "real" mathematics consists of basic computational facts like $2+2=4$, and the rest of mathematics is merely "ideal," with an instrumental value for deriving computational results, but no direct computational meaning. My late colleague Dick Jeffrey proposed instead that one should think of mathematics as being logical in the same sense in which physics is empirical: the *data* of mathematics are logical, as those of physics are empirical, though there can be no question of defining all mathematical notions in strictly logical terms, or all physical notions in strictly empirical ones. Mathematics becomes, on this view, a giant engine for generating logical results, as physics is a giant engine for generating empirical results. Hilbert's proportion is modified by Jeffrey so that it reads thus:

logical : mathematics :: empirical : physics.

The connection between the Jeffrey idea and the Heck idea is that predicative logicism provides enough mathematics to connect the basic computational facts that figure in the Hilbert proportion with the logical facts that figure in the Jeffrey proportion. So, though predicativist logicism falls far short of the whole of mathematics it would be possible to regard it as providing the *data* for mathematics.

The question I raise about all this is whether the engine is doing any real work: do sophisticated mathematical theories (such as standard Zermelo–Frankel set theory or the proposed Boolos–Bernays set theory) actually make available any more logical "predictions" in the way sophisticated theories in physics make available more empirical predictions? I show how (and in what sense) certain results of Julia Robinson and Yuri Matiyasevich in mathematical logic yield an affirmative answer to this question. (More technical details are provided in a forthcoming paper "Protocol Sentences for Lite Logicism.")

MODELS AND MEANING

"Tarski's tort" (the other previously unpublished item in the collection) is a sermon on the evils of confusing, under the label "semantics," a formal or mathematical theory of models with a linguistic or philosophical theory of meaning. Tarski's infringing on the linguists' trademark "semantics," and transferring it from the theory of meaning to the theory of models, encourages such confusion, which has several potentially bad consequences.

Such a confusion may lead, on the one hand, to erroneous suspicions that many ordinary locutions involve covert existential assumptions about dubious entities. (For example, one may fall into a fallacy of equivocation and argue that since possible worlds are present in the model theory of modal logic, they are therefore present in the semantics of modality, and are therefore present in the meaning of modal locutions.) Such a confusion may lead, on the other hand, to unwarranted complacency about the meaningfulness of dubious notions. (For example, one may fall into a different fallacy of equivocation and argue that since quantified modal logic has a rigorous model theory, it therefore has a rigorous semantics, and therefore has a rigorous meaning.)

Confusion of models and meaning under the label "semantics" may also give undeserved initial credibility to truth-conditional theories of meaning, through their mistaken association with the prestigious name of Tarski. One point on which I think Dummett is entirely right is rejection of the truth-conditional theory of meaning, and insistence that meaning must be explained in terms of rules of use, though my reasons for holding this view are rather different from Dummett's. (For me, perhaps the most incredible feature of the truth-conditional theory is the assumption that truth is an innate idea, possession of which is a prerequisite for all language-learning. I find much more plausible the suggestion that the idea of truth is acquired at the time the word "true" is acquired, and that acquisition of the idea consists in internalizing certain rules for the use of the word.)

Such a view leads naturally to the "inconsistency theory" of truth advocated in different forms by my teacher Charles Chihara, my student John Barker, and (not under any influence of mine) my son Alexi Burgess. There being by Church's theorem no effective test for the inconsistency of rules, it would be a miracle if all the rules we ever internalized were consistent. The simplest and most natural rules for the use of "true," permitting inference back and forth between *p* and *it is true that p*, are inconsistent. Acceptance that these inconsistent rules of use are the ones we internalize when we acquire the word "true" and therewith the idea of truth provides the simplest and most natural explanation of the intractability of the liar and related paradoxes, an explanation favored not only by the persons already mentioned, but by Tarski himself. All these issues are touched on in "Tarski's tort," though none is argued in all the detail it deserves. (Some of the issues were previously aired in Burgess (2002b) and elsewhere.) The warnings about fallacies of equivocation are pertinent to other papers in the second part of the collection, which is one reason for placing this paper first in that part.

MODELS AND MODALITY

Nothing is more important in approaching modal logic than to bear constantly in mind the distinction between two kinds of necessity: metaphysical necessity – what could not have been otherwise – and logical necessity – what it would be self-contradictory to deny. Modal logic has been characterized from early on by a great proliferation of systems, even at the level of sentential logic. In a way this is all to the good, since different conceptions of modality, once one has learned to distinguish them, may call for different formal systems. But we do need to distinguish the different notions before we can meaningfully ask which system is right for which notion. And though philosophers and logicians nowadays are more aware of the distinctions among various kinds of modality than they were formerly, the problem of determining which formal system is appropriate for which conception of modality is one that still has received surprisingly little attention.

In "Which modal logic is the right one?" (1999) I take up this question for the case of the original conception of necessity of C. I. Lewis, the founder of modern modal logic, for whom necessity was logical. And the first thing that needs to be said about logical modality is that it comes in two distinguishable kinds, a "semantic" or model-theoretic notion of *validity* and a "syntactic" or proof-theoretic notion of *demonstrability*. (For first-order logic demonstrability and validity coincide in extension, by the Gödel completeness theorem, but they are still conceptually distinct; for other logics they need not coincide even in extension.) The former makes necessity a matter of *truth* by virtue of logical form alone, the latter a matter of *verifiability* by means of logical methods alone. The common conjecture is that the system known as S5 is the correct logic for the former, and that known as S4 for the latter. The well-known Kripke model theory for modal logic is a useful tool, but hardly in itself provides a complete proof of either conjecture. (As the originator of this model theory said, "There is no mathematical substitute for philosophy.") As it happens, the conjecture about S5 admits a fairly easy proof, which I expound, while for the conjecture about S4 only partial results are available, which I explore.

Turning from logical to metaphysical necessity, no single tool is more useful for understanding the logical aspects of the latter than the analogy between mood and modal logic on the one hand, and tense and temporal logic on the other. One of the older puzzles about metaphysical modality is Frederic Fitch's paradox of knowability, which purports to demonstrate the incoherence of the view that anything that is true could be known. "Can truth out?" (2006) first considers what light can be shed on this puzzle by looking at a temporal analogue, and then by applying Arthur Prior's branching-futures logic, in which modal and temporal elements are combined. As with the previous paper, a certain amount of progress is possible, but a complete solution remains elusive.

MODALITY AND REFERENCE

Logical necessity was originally what modal logicians had generally meant the box symbol to represent, and I believe that Quine was entirely correct in asserting that with *that* understanding of the symbol, quantifying into modal contexts makes no sense. Quine's complaint was that to make non-trivial sense of $\exists x \Box Fx$ one must make sense of so-called *de re* modality, of the notion of an open sentence *Fx* being necessarily true *of a thing*, and this is impossible (or at any rate, has not been done by the proponents of quantified modal logic) if "necessarily true" is to mean "true by virtue of meaning." For a *thing*, as opposed to an expression denoting a thing, has not got a meaning for anything to be true by virtue of. Truth by virtue of meaning is an inherently *de dicto* notion, applicable to closed sentences. Quine underscored his point by illustrating the difficulty of reducing *de re* to *de dicto* modality. One can't say $\Box Fx$ is true of the object *b* if and only if $\Box Ft$ is true, where *t* is a term denoting *b*, because $\Box Ft$ may be true for some terms denoting the object and false for other terms denoting the same object.

The early response of modal logicians to Quine's critique, by which I mean the responses prior to Kripke's "Naming and necessity" (1972), involved an appeal, not to a distinction between metaphysical and logical modality, but rather to (purely formal "semantics" and/or to) the magical properties of Russellian logically proper names. In simplest terms, the "solution" would be that $\Box Fx$ is true of *b* if and only if $\Box Fn$ is true, where *n* is a "name" of *b*, it being assumed that if $\Box Fn$ is true for one name it will be true for all. I believe this line of response is a total failure, and anyone acquainted with the views of Mill ought to have been aware that it must fail. For it will be recalled that Mill, before Russell, held that *a name has only a denotation, not a connotation.* He also held, like the modal logicians who identified □ with truth by virtue of meaning, that *all necessity is verbal necessity*, deriving from relations among the connotations of words. What those responding to Quine ought to have remembered is that, having committed himself to those two views, he inevitably found himself committed to a third view, that *there are no individual essences*: "all *G*s are *F*s" may be necessarily true because being an F may be part of the connotation of G, but "*n* is an *F*" cannot be necessarily true, because *n* has no connotation for being an *F* to be part of. Positing Millian/Russellian "names" may permit the reduction of *de re* to *de dicto* modality, where the relevant *dicta* involve such names, but only at the cost of depriving *de dicto* modality, where again the *dicta* involve such names, of any sense – so long as one continues to read □ as truth by virtue of meaning.

What is now the fashionable view evaluates the early responses to Quine much more positively than I do, when it does not outright read Kripke's ideas back into earlier texts. That Quine was right as against his early critics is the view that I, going against fashion, defend in "*Quinus ab omni naevo vindicatus*" (1998). The origin of the curious title is explained in the article.

Closely linked with the issue of metaphysical versus logical necessity is the question of the status of identities linking proper names, as in "Hesperus is Phosphorus." This question has been the topic of an immense body of literature in philosophy of language. Like Kripke, I on the one hand reject descriptivist theories of proper names, but on the other hand equally reject "direct reference" theories. I am attracted to a third view based on distinguishing two senses of "sense," mode of presentation versus descriptive content, a view rather tentatively (and certainly non-polemically) put forward in connection with Kripke's Puzzling Pierre problem in "Translating names" (2005).

But returning for a moment to "*Quinus*," since some readers like nothing better than polemics between academics, and others like nothing less, all potential readers should be informed in advance that "*Quinus*" is as polemical as anything I have ever written (though much of the polemic is relegated to footnotes), and several degrees more so than anything else in this collection. Even the explanation *why* the paper is polemical must itself inevitably be somewhat polemical, and so I will relegate it to a parenthetical paragraph which those averse to polemics may skip, along with the paper itself.

(My paper is explicitly a response to a paper by Ruth Barcan Marcus from the early 1960s, but it is also implicitly a response to a widely circulated letter by the same writer from the middle 1980s, discussed in the editorial introduction to Humphreys and Fetzer (1998). This letter is the original source for the claim that Kripke's ideas were taken without acknowledgment from the early Marcus paper. The response to such allegations is that Kripke could not have stolen his ideas from the indicated source, since neither those important and original contributions nor any others were present there to be plagiarized. In my paper I do not mince words in presenting this defense. Marcus's insinuations are more directly addressed in my paper (Burgess 1996). The contents of her letter, with elaborations but without acknowledgment of that letter as a source, reappear in the work of one of its many recipients, Quentin Smith. His version is addressed in my paper (Burgess 1998a).)

HERMENEUTIC CRITICISM OF CLASSICAL LOGIC: RELEVANTISM

In Burgess (1992) I offered qualified defense of classical logic, leaving plenty of room from additions and amendments once one moves beyond the realm of mathematics. I introduced in the paper a distinction that seems to me crucial in evaluating certain criticisms of classical logic, namely, the distinction between *prescriptive* criticism, according to which classical logicians have correctly described the incorrect logical practices of classical mathematicians, and *descriptive* criticism, which maintains that classical logicians have incorrectly described the correct logical practices of classical mathematicians. The distinction parallels that between revolutionary and hermeneutic nominalism. The remaining two items in the collection discuss examples of the two types of anti-classical logics. (Both have modal aspects or admit modal interpretations, and certainly can be studied by some of the methods, notably Kripke models, developed for modal logic, and insofar as this is so may be squeezed under the "models and modality" heading for the second part of the collection.)

Much descriptive criticism of classical logic, especially that from the old ordinary language school, was essentially anti-formal. The best example of descriptive criticism in the service of a rival formal logic was provided by the "relevance" logic of A. R. Anderson and Nuel D. Belnap, Jr. in its original form, back when it was a more or less unified philosophical school of thought, denouncing and deriding classical logic, and recommending one or the other of two specific candidates, the systems E and R, as replacement. The enterprise has since been renamed "relevant" logic and devolved into the study of a loose collection of formal systems linked by family resemblances, with quite varied intended or suggested applications.

"Relevance: a fallacy?" (1992) was devoted to presenting counterexamples to one key early claim of relevance logic, never up to the time I was writing explicitly retracted in print. Relevantists rejected, as "a simple inferential mistake," the inference from A $\vee$ B and $\sim$A to B. But inference from "A or B" and "not A" to B is common in classical mathematics and elsewhere. Anderson gave little attention to the resulting tension, but Belnap attempted to resolve it by claiming that "or" in ordinary language generally means not the extensional $\vee$ but some intensional $+$. It was this *specific* claim I challenged. I do not have a globally negative view of the relevantist enterprise. Not only has there been some impressive technical work by Saul Kripke, Kit Fine, Alasdair Urquhart, Harvey Friedman, and others, but the "first degree" and "pure implicational" fragments of several relevantist systems do have coherent motivations, as do various related logics, such as Neil Tennant's idiosyncratic version of relevantism or "logical perfectionism." I do not, however, think either of the main relevantist systems E or R *as a whole* has any coherent motivation, and more importantly, I do not think there was any merit in the original relevantist criticism and caricature of classical logic.

My paper prompted replies in the same journal by Chris Mortensen and Stephen Read, to which I in turn responded, again in that journal. Ironically, though members of a group who pride themselves on their sense of "relevance," the two writers who directly replied to my paper simply could not confine themselves to addressing the specific issue I had treated, but insisted on offering, as if this somehow refuted my claims, expositions of motivations for relevantism quite different from those of Anderson and Belnap (and quite different from each other). This led me to write, "The champion of classical logic faces in relevantism not a dragon but a hydra." It also led me to reiterate my original point and present a variety of further counterexamples, drawn from several sources. The polemics do not seem to me worth reprinting, but I do refer any reader not convinced by the two examples in my original paper to the first list of examples (borrowed from authors ranging from E. M. Curley to Saul Kripke) in §2 of the first of my replies to critics (Burgess 1983c).

My second reply (Burgess 1984b) contained one more example: by the regulations of a certain government agency, a citizen *C* is entitled to a pension if and only if *C either* satisfies certain age requirements or satisfies certain disability requirements. An employee *Z* of the agency is presented with documents establishing that *C* is disabled. *Z* transmits to fellow employee *Y* the information that *C* is entitled to a pension (i.e. is either aged or disabled). *Y* subsequently receives from another source the information that *C* is not aged, and concludes that *C* must be disabled.

REVOLUTIONARY CRITICISM OF CLASSICAL LOGIC: INTUITIONISM

While it may be contentious whether relevantism really does provide an example of *descriptive* criticism of classical logic, it is beyond controversy that intuitionism provides an example of *prescriptive* criticism. Adherence to intuitionistic logic would certainly require major reforms in mathematics. Very likely acceptance of the verificationist concerns that motivate contemporary intuitionism would require still more dramatic reforms of a nature we cannot yet quite take in, in empirical science. For in the empirical realm we have to contend with two phenomena that do not arise in mathematics: first, we generally have to deal not with apodictic proof but with defeasible presumption; second, it may happen that though each of two assertions may be potentially empirically testable, performing the operations needed to test one may preclude performing the operations needed to test the other. (A DNA sample, for instance, may be entirely used up by whichever of the two tests we choose to perform first; nothing analogous ever happens with operations on numbers.) The ultimate verificationist logic may well have to combine features of intuitionistic, nonmonotonic, and quantum logics.

Michael Dummett is perhaps the single most influential representative of *anti*-naturalism in contemporary philosophy, though his contributions extend far beyond that role. (He is, among many other things, the leader of Frege studies, and in that capacity motivated, among many other things, the exploration of the predicativist variant of Frege's theory discussed in the last paper of Part I.) His paper (Dummett 1973a) inaugurated a new era in the classical-intuitionist debate over logic and mathematics, and was the font from which what is now a vast stream of "anti-realist" literature first sprang. A noteworthy feature of Dummett's approach is that, like Brouwer but unlike almost every writer on intuitionism in-between, he takes the considerations that motivate intuitionism to apply not just to mathematics but to all areas of discourse. Mathematics is special only in that we have a better idea of what a revision of present practice would amount to in that area than in any other.

For that reason I have placed "Dummett's case for intuitionism" in this part of the collection, rather than the part on philosophy of mathematics specifically. But to repeat what I said in the introduction to Part I, the division of the collection into parts is not to be taken too seriously.

"Dummett's case" advances two kinds of countercriticisms of the criticism of classical logic that appears in Dummett (1973a). In the first part of the paper, making an explicit mention of Noam Chomsky and an implicit allusion to John Searle, I object to Dummett's unargued behaviorist assumptions. In this part of the paper I am arguing as devil's advocate, since I do in the end agree with Dummett in rejecting truth-conditional theories of meaning. Some followers of Dummett have objected to the label "behaviorist"; but I think this is largely a terminological issue. If I say that Sextus, Cicero, Montaigne, Bayle, and Hume were all skeptics, I do not imply that their views were identical; likewise if I say that Watson, Skinner, Quine, Ryle, and Dummett are all behaviorists. It is clear that Dummett's position, however one labels it, remains light-years away from Searle's, let alone Chomsky's.

What I object to in the second half of the paper is the lack of explicitness about how the transition from *is* to *ought*, from the premise that a truth-conditional theory of meaning *is* incorrect for classical mathematics to the conclusion that classical mathematics *ought to be* revised, is supposed to be made. Hints are thrown out, to be sure, which have been developed in different ways by Dummett himself in later works (especially *The Logical Basis of Metaphysics*) and in more formally terms by Dag Prawitz, Niel Tennant, and others. The most explicit proposals suggest that the meanings of logical operators should be thought of as constituted by something like the introduction and/or elimination rules for those operators in a natural deduction system. Then intuitionistic logic is claimed to be better than classical logic because there is a better "balance" between its introduction and elimination rules.

In "Dummett's case" I largely confine myself to saying that this aesthetic benefit of "balance" could hardly outweigh considerations related to the needs of applications, objecting to Dummett's apparent indifference to the latter in advocating revision of mathematics. How much of the standard mathematical curriculum for physicists or engineers can be developed within this or that restrictive finitist, constructivist, predicativist, or whatever framework is a topic that has been fairly intensively investigated by logicians (though to this day I know of no serious study of how a finitist, constructivist, or predicativist should interpret *mixed* contexts involving both mathematical and physical objects and properties, something nominalists by contrast take to be of central concern). Dummett simply omits to address this issue at all in his key paper, and that omission would, I think, raise the eyebrows of any logician.

I have recently written a sequel to "Dummett's case" (Burgess 2005e), but on the advice of an anonymous referee have decided not to include it here, as its tone may in places offend Dummettians (even more so than that of my older paper, if that is possible). The main point of the paper still seems to me worth making: neither the view that the meaning of each logical particle is constituted by its introduction rule (for example, the rule "from *A* and *B* to infer *A* & *B*"), nor the view that it is constituted by the introduction rule together with the corresponding elimination rule (for example, "from *A* & *B* to infer *A* and to infer *B*") is tenable if "meaning" is supposed to be *what guides use*. For we constantly "introduce" and "eliminate" conjunctions, say, by quite other means than conjunction introduction and elimination (for instance, we may arrive at a conjunction by universal instantiation and by *modus ponens*, which is to say, by the elimination rules for the universal quantifier and for the conditional). And these steps cannot be justified from the point of view of one who really, truly, and sincerely takes the sole direct guide to the use of "&" to be the introduction and/or elimination rules for that connective. Nor – and this is the crucial point of the paper – can any metatheorem provide an indirect justification, since the proof of the metatheorem inevitably involves introducing and eliminating conjunctions according to rules that, at least until the proof of the metatheorem is complete, have not yet been justified from the point of view in question. (Moreover, the application of any metatheorem to any particular case would anyhow require universal instantiation and *modus ponens*.)

More generally, my paper points out how frequently writers (including not only neo-intuitionists, but relevantists and nominalists) who profess to reject certain principles of classical logic (or mathematics), and appeal to metatheorems supposed to show that nothing much is lost thereby, can be caught using the supposedly rejected principles in the proofs of those very metatheorems. To my mind, this is one striking illustration of how difficult it is to be a *genuine* dissenter from classical logic and mathematics.

PART I

Mathematics

I

Numbers and ideas

I REALISM VS NOMINALISM

Philosophy is a subject in which there is very little agreement. This is so almost by definition, for if it happens that in some area of philosophy inquirers begin to achieve stable agreement about some substantial range of issues, straightaway one ceases to think of that area as part of "philosophy," and begins to call it something else. This happened with physics or "natural philosophy" in the seventeenth century, and has happened with any number of other disciplines in the centuries since. Philosophy is left with whatever remains a matter of doubt and dispute.

Philosophy of mathematics, in particular, is an area where there are very profound disagreements. In this respect philosophy of mathematics is radically unlike mathematics itself, where there are today scarcely ever any controversies over the correctness of important results, once published in refereed journals. Some professional mathematicians are also amateur philosophers, and the best way for an observer to guess whether such persons are talking mathematics or philosophy on a given occasion is to look whether they are agreeing or disagreeing.

One major issue dividing philosophers of mathematics is that of the nature and existence of mathematical objects and entities, such as *numbers*, by which I will always mean *positive integers* 1, 2, 3, and so on. The problem arises because, though it is common to contrast matter and mind as if the two exhausted the possibilities, numbers do not fit comfortably into either the material or the mental category.

Clearly numbers are not material bodies. The so-called numbers on the front of a house, marking its street address, may indeed be made of brass or wood or plastic. But these "numbers" are not the numbers we speak of when we say that two is an even number, or that three is an odd number, or that both are prime numbers. Rather, they are *numerals*, or names of numbers.

Almost equally clearly, numbers are not mental in the way that, say, dreams or headaches are. They are not private to an individual. One does not speak of my number two and your number two, his number two and her number two, but simply of *the* number two. The individual, say a school child doing a simple sum, experiences the numbers as something external, about which he or she is *not* free to think whatever he or she wants.

But if numbers are not material bodies or private experiences, what (if anything) are they? Among professional academic philosophers, which is to say university professors of the subject, the most commonly held views are two, for want of better terms called *realism* and *nominalism*.

Realism maintains that numbers exist, and are of a very different nature from human ideas: indeed, they differ quite as much from human ideas as they do from material bodies. They are *abstract* entities, to which it makes no sense to ascribe a position in space or date in time, and which are not causally active or acted upon. There is nowhere to go to look for a number, and you cannot do anything to a number, any more than a number can do anything to you.

Nominalism maintains that numbers do not exist, and that theorems of mathematics asserting the existence of numbers are untrue, just like fairy tales asserting the existence of gnomes. To be sure, much of mathematics is applicable in science and everyday life in a way that fairy tales generally are not, but that, according to nominalists, only shows it is a *useful* fiction, not that it is non-fiction.

There are problems for both opposing philosophical views, and the problems of each are cited by the adherents of the other as reasons for embracing *it* instead. And formerly there were among philosophers also many who maintained a third view, *conceptualism* or *idealism*, according to which numbers exist, but only as shared human concepts or ideas.

The view has traditionally been popular among anthropologists and other social scientists, whose special subject matter is precisely the shared ideas of a culture. They point out that taking numbers to be such shared or *communal* ideas sufficiently explains why the school child doing a simple sum does not feel free to make up an answer at will. If numbers are ideas shared by a culture, no one member of that culture has the authority to change the rules of addition, any more than to change the rules of grammar of the culture's language.

The anthropological view has also found adherents among mathematics educators. Rather more surprisingly, the same view has won adherents among the minority of professional mathematicians who are also amateur philosophers.[1]

[1] The classical expression of the anthropological view is that of White (1947). For a recent endorsement by a mathematician, see Hersh (1997), a book that makes a professional philosopher's hair stand on end.

Conceptualist and idealist views, however, were subjected along with other nineteenth-century views to a scathing critique by the late nineteenth-century German mathematician and philosopher Gottlob Frege.[2] Largely as a result of that critique, the anthropological view today has virtually no adherents among professional academic philosophers. Its rejection is one of the rare cases of general agreement and consensus on an issue in philosophy.

Precisely because there is such general agreement, philosophers seldom stop to explain, in language more modern than Frege's, just what is wrong with the view that so many anthropologists, sociologists, psychologists, mathematics educators, and even mathematicians have found attractive. It is this task of explanation that I will be undertaking in the present essay, using an example of a kind that definitely would *not* have been used by Frege.

2 BIGFOOT

Let us begin by considering the proposition that *Bigfoot*, also known as the *Sasquatch* – a cousin of the *Abominable Snowman* or *Yeti* – exists in the realm of shared human ideas and concepts. Now certainly there is *something* in the neighborhood that exists in the realm of shared human ideas and concepts, namely, the shared human idea or concept of Bigfoot. This is the idea of a large, hairy, humanoid creature inhabiting the wilder parts of the Pacific Northwest, from northern California to British Columbia.

There are even people who claim to have sighted individual Bigfeet, and to have formed ideas of these individuals, even to the point of giving them names like "Harry" or "Harriet." The idea of an individual Bigfoot includes the traits that are common to all Bigfeet according to the general idea of Bigfoot, but also more specific elements: for instance, Harry is male and Harriet is female. These ideas of individual Bigfeet are less widely shared than the idea of the species, but we may suppose they are at least shared among members of the International Society for Cryptozoology, who take a special interest in such things.

The majority view among zoologists is that there do not, in fact, exist any large, hairy, humanoid creatures, and that the alleged sightings of Harry, Harriet, and other individual Bigfeet were either illusions or hoaxes. But I ask you to join me in assuming, just for the moment, that the majority is wrong, and that creatures of the kind indicated, including Harry and Harriet, do exist. On this assumption, I will argue, two things should be clear.

[2] See Frege (1884), English translation by Austin (1960). The critical portions (the part of the book relevant to the present essay) are reprinted in Benacerraf and Putnam (1983).

The first is that Harry, Harriet, and other large, hairy, humanoid creatures inhabiting the wilder parts of the Pacific Northwest are very different sorts of things from shared human ideas and concepts, and in particular are very different sorts of things from the ideas and concepts of Harry, of Harriet, and of Bigfoot in general. They differ in absolutely fundamental respects, for instance, in their location in space and time.

Let us consider space, for instance. (Similar considerations would apply to time.) It is not clear whether or where a shared human idea or concept should be thought of as located in space, but presumably if it is located anywhere, it is located where the human beings who share it are located. Thus if the International Society for Cryptozoology holds its annual convention on the banks of Loch Ness, the idea of Bigfoot in general, and the ideas of Harry and Harriet in particular, are located mainly in Scotland. Harry, Harriet, and the rest of their kind, however, are still located in Washington or Oregon or thereabouts. The creatures cannot be the ideas, because the two are located in different places.

The creatures differ from the ideas also in respect of how many of them there are. People have ideas of Harry, Harriet, and several more Bigfeet that have allegedly come into contact with human beings; but there are supposed to be, according to the minority view I have asked you to assume for the moment, more Bigfeet than just these: more individuals like Harry and Harriet than there are shared human ideas of individual Bigfeet. So again the creatures cannot be the ideas, since there are more of the former than of the latter.

A second point I hope will be clear is that it is the flesh-and-blood creatures, not the ideas, that are the Bigfeet. The term "Bigfoot" refers to the inhabitants of the wilds of Washington and Oregon, not to the contents of the minds or brains of the cryptozoologists assembled in Scotland. If we wish to refer to the latter, we must use some other expression than the word "Bigfoot," such as the phrase "the idea of Bigfoot."

In short, on the minority view, according to which the flesh-and-blood creatures do exist, the following is the case: Bigfeet, being flesh-and-blood creatures, are not ideas, and are more numerous than the ideas of them and located in a different place from those ideas.

3 NUMBERS

Are things any different on the majority view? It is when one assumes that there are no such flesh-and-blood creatures that some are tempted to say that the Bigfoot in general, or Harry and Harriet in particular, are human ideas. I think this temptation should be rejected.

Let me say straightaway that it would be pointless to object to someone expressing *disbelief* in Bigfoot by saying, "Bigfoot exists only in the imagination of the credulous," or something of the sort. Someone might well say this – I might well say it myself, for that matter, when not talking philosophy – and mean it only as a manner of speaking, as a way of saying "Bigfoot doesn't exist at all, though some credulous persons imagine that it does." The proposition I want to consider, however, is that Bigfoot *literally* does exist, but only in the realm of shared human ideas and concepts, where, according to the anthropological view, numbers also have their being.

To indicate the reasons why I reject this proposition, suppose the population of some endangered forest or swamp species falls until there is only one left. So long as this one surviving flesh-and-blood or wood-and-sap organism lives, considerations of the kind already adduced in the case of Bigfoot indicate that *it* is the only member of the species, and *it* is not an idea, from which it follows that the members of the species are not ideas.

Now suppose this last survivor also perishes. Are we now to say that the species still has members, but that the members of the species are now ideas? Should we say that the species has not become extinct but rather has undergone a *metamorphosis*, transcending its former carnal or xyline nature, and taken on a conceptual essence: that its members have cast aside their fleshly or wooden bodies, and are now made of whatever ideas are made of? Should we say that the species has undertaken a *migration*, abandoning the woods or marshes that were once its home, and occupying now instead a niche in the minds or brains of human subjects?

It seems to me about as plain as anything can be in philosophy – where admittedly things are never as plain as they are in some other disciplines – that this is *not* what we should say, and that the *correct* way to describe the situation is by saying that creatures of this animal or plant species *simply no longer exist at all*, though of course human ideas about them do exist, and may perhaps continue to exist as long as the human species does.

Likewise in the case of Bigfoot. If the forest creature exists, then Bigfoot is that forest creature, and is something very different from an idea. If the forest creature does not exist, then Bigfoot is, so to speak, even *more* different from an idea: for in that case Bigfoot is *nothing*, while the idea is at least *something*, and what could be more different than something and nothing?

The case is the same, I maintain, with our shared human ideas and concept of number in general, and of individual numbers such as one or two or three. (Again the individual ideas contain whatever is contained in the general idea, plus additional distinguishing elements. We no longer imagine, as did the Pythagoreans, that two is female and three is male, but, for instance, two is even and three is odd.)

These ideas are clear enough, I maintain, to indicate that one, two, three, and the other numbers, if they exist at all, do not have the same sort of spatial or temporal features as human ideas, and above all are more numerous than human ideas could possibly be.

Taking first issues of time and place, mathematics is used throughout science, and mathematical objects and entities are referred to in all its branches, including those like cosmology that deal with times and places very remote from any inhabited by human beings. Are we to say that a cosmologist's estimates of the relative numbers of heavy and light elements at a certain stage in the early evolution of the universe must be wrong, because there were no numbers at all back then, no human beings having yet evolved to create them? Surely not.

And then there is the matter of infinity. It is a crucial feature of the concept of the number system that it has infinitely many elements, that there are infinitely many numbers. But surely human beings have formed ideas or concepts of only finitely many of them. There simply are not enough human ideas and concepts for each number to be one. *Some* numbers at least must therefore either enjoy a mode of existence different from that of any human idea, as realists maintain, or else must simply fail to exist, as nominalists hold. And is it not preposterous to maintain that while one of the pair realism or nominalism gives the correct account of mathematical existence in the case of *some* numbers, conceptualism is correct *for the rest*? Surely the question of the existence and nature of numbers has a *uniform* answer, and if conceptualism fails in *any* case, then it must fail in *all*.

4 REALISM VS NOMINALISM REVISITED

Such, then, are some of the principal reasons why I and almost all professional philosophers of mathematics reject conceptualism, and consider the only real issue to be that between nominalism and realism. This last issue is far too large to be thrashed out here, but I do wish to say a word about it, and in particular about the character of the *realist* position, which very often tends to be misrepresented. Nominalists do not believe in numbers because they cannot *see* them (or see any visible effects *caused* by them), and tend to represent their opponents as claiming that they *can* see them.

According to an old story, Plato was once lecturing in his Academy on his Forms, and was speaking of the forms of "tableness" and "cupness." Diogenes the Cynic interrupted and said, "O Plato, I see the table and the cup, but the tableness and the cupness I do not see." To this Plato replied, "Very naturally, Diogenes, since you have eyes, by which material things are perceived, but lack Intellect, by which the forms are seen."[3]

Nominalists tend to represent their opponents as Platonists, maintaining that if numbers do not emit electromagnetic radiation to which the eye is sensitive, then they must be emitting something else, perhaps *noetic rays*, which can be sensed by some other organ, perhaps the *pineal gland*. This, however, is a misrepresentation of realism. Or at least, I have never known a single realist who was in any meaningful sense a Platonist.

What is actually the case is that anti-nominalists take much more seriously than nominalists the thought that *mathematics* is a human creation, since mathematics is a body of theory expressed in language, and *language* is a human creation.

Now creating a language involves creating certain rules for its use. Among these is, I believe, a rule to the effect that tense and date are not to be applied to mathematical existence assertions. One can say "There exist infinitely many prime numbers," but to ask "How many of them already existed in 1000 BCE, or during the Cenozoic Era?" is to commit a kind of grammatical solecism.

Nominalists say they are opposed to the view that numbers are "eternal," existing "outside of time." But to say that numbers are "eternal" is a misleadingly Platonistic way of putting the simple negative grammatical fact of the inapplicability of tense distinctions in mathematical contexts. That simple grammatical point is all the realist really believes about the "timelessness" of number.

(By contrast with the case of the numbers themselves, it makes perfect sense to ask whether the *idea* or *concept* of prime number had emerged by 1000 BCE – the issue involved would be that of the interpretation of certain Babylonian tablets and Egyptian papyri – and it makes perfect sense to assert that it had *not* emerged in the Age of the Dinosaurs. This difference between the "timeless" numbers proper and datable ideas of them was one of the points I was arguing in rejecting conceptualism.)

Likewise, there are certain rules or standards as to what counts as adequate or sufficient to establish or prove a mathematical existence theorem, and by these rules Euclid's theorem on the existence of infinitely many prime numbers is as well-established as anything can be.

The *nominalists* assume that they have an understanding of what it would be for a mathematical object or entity to exist that is independent

[3] See the life of Diogenes the Cynic in Diogenes Laertius (1925).

of ordinary mathematical standards of sufficient proof, by reference to which understanding they can *criticize* the ordinary mathematical standards. So-called realism is really just *skepticism* about the existence of any understanding of what "existence" means in mathematics that is independent of ordinary mathematical standards for evaluating existence proofs. The nominalist denies the existence of numbers, while the realist denies that the nominalist understands what is meant by "existence" as applied to numbers.

Thus the realists think the nominalists are confused. But realists and nominalists agree that the conceptualists are confused, and while I cannot hope to have convinced anyone by the foregoing very brief remarks that the realists are right as against the nominalists, I hope I have convinced some of you that realists and nominalists are right in their common opposition to conceptualism.

2

Why I am not a nominalist

INTRODUCTION

The sum of the divisors of 220 is 284, and the sum of the divisors of 284 is 220. The Pythagoreans spoke of numbers so related as being *amicable*. I do not know how this ancient teaching should be taken, but surely nobody nowadays, except perhaps a stray numerologist or two, would imagine that numbers are literally capable of forming friendships. A number is just not the sort of thing that can enjoy a social life. And this is but the least of a number's lacks.

A number lacks a position in space, such as tables, chairs, and other material bodies possess. It lacks dates in time, such as dreams, headaches, and other contents of minds possess. It lacks all visible, tangible, audible properties. In a word, it is *abstract*.

Disbelievers in numbers and other abstract entities or "universals" have come to be called *nominalists*. Nominalism has always attracted philosophers of the hard-headed, no-nonsense type. But does it not conflict with modern science, which speaks the language of abstract mathematics?

I INSTRUMENTALIST NOMINALISM

Some nominalists concede that their philosophy of mathematics conflicts with science by implying that science, when it speaks the language of mathematics, is not speaking truly. These nominalists adopt an *instrumentalist* philosophy of science, according to which science is just a useful mythology, and no sort of approximation to or idealization of the truth. Truth is to be sought, rather, in a philosophy prior and superior to science.

The position of the well-known nominalist Nelson Goodman is best understood as a subtle and sophisticated variation on instrumentalism. For Goodman, science is less a useful fiction than useful *nonsense*. But whereas a straightforward, simple-minded instrumentalist would be willing to label science as untrue and let it go at that, Goodman holds that the philosopher ought at least to attempt to give some sense to the scientist's otherwise senseless productions by reconstructing them nominalistically:

> The nominalist does not presume to restrict the scientist. The scientist may use Platonistic class constructions, complex numbers, divination by inspection of entrails, or any claptrappery that he thinks may help him get the results he wants. But what he produces then becomes raw material for the philosopher, whose task is to make sense of all this: to clarify, simplify, explain, interpret in understandable terms ... Nominalism is a restraint the philosopher imposes on himself, just because he feels he cannot otherwise make real sense of what is put before him. (Goodman 1956, objection vii)

Goodman's own steps towards nominalistic reconstruction of science (taken jointly with Quine in Goodman and Quine (1947)) never led very far. So presumably for Goodman the bulk of science remains nonsensical.

Most recent philosophers of science, even those nominalistically inclined, have been hostile toward instrumentalist philosophies like Goodman's for a couple of good reasons. For one thing, since science is just an outgrowth of common sense, there can be no sharp dividing line between them. The most abstruse theoretical physics is connected in a thousand ways through experimental and applied science, through engineering and technology, to everyday belief. And much of everyday belief is couched in the vocabulary of mathematics, albeit of a sort more elementary than that which figures in general relativity theory or quantum mechanics. The philosopher who begins by rejecting theoretical physics as fiction will find no logical place to stop, and in the end will be unable, without inconsistency and self-contradiction, to accept commonsense belief as fact.

For another thing, the behavior of instrumentalists when not consciously philosophizing strongly suggests that their professed disbelief in science is a sham. Catch them off guard, and you are likely to find them classing the Steady State theory as false, and the Big Bang theory as true, just like the rest of us. The instrumentalist seems to be "engaging in intellectual double-think: taking back in [his] scientific moments what [he] asserts in doing science" (Field 1980, p. 2). He seems to be "an irrational person ... who is unwilling to accept the consequences of his own theories" (Chihara 1973, p. 63).

It is on account of such *slippery slope* and *insincerity* objections that instrumentalism is not a live option for most contemporary nominalists; and it is certainly not a live option for me.

2 SCIENTIFIC DISPENSABILITY AND NONEXISTENCE

Some anti-nominalists have argued that the conflict between nominalism and science is so strong that nothing like modern science as we know it could survive if the nominalist ban on mathematical abstractions were accepted. Such a position has been reluctantly maintained by the ex-nominalist Quine ever since the failure of his joint attempt with Goodman at nominalistic reconstruction. Such a position was also maintained, under Quine's influence, by Hilary Putnam, during his phase of enthusiastic realism.

> I have explained early and late that I see no way of meeting the needs of scientific theory ... without admitting universals irreducibly into our ontology ... Nominalism ... is evidently inadequate to a modern scientific system of the world. (Quine 1981, pp. 182–3)

> It has been repeatedly pointed out that such a [nominalistic] language is inadequate for the purposes of science ... The restrictions of nominalism are devastating ... It is not just "mathematics" but physics as well that we would have to give up. (Putnam 1971, p. 35)

In short, Quine and Putnam have maintained that mathematical objects are *scientifically indispensable.*

The refutation of this thesis has been the first aim of the most prominent recent nominalist writers, Charles Chihara and Hartry Field. The programs of nominalistic reconstruction developed in their books (Chihara 1973; Field 1980) are reviewed in outline in the Appendix to the present chapter. Suffice it to say here that Chihara and Field draw on results from advanced research in the foundations of mathematics (predicative analysis, measurement theory, proof theory), and that Chihara assigns the work normally done by mathematical abstractions to certain *modal* notions (including that of the possibility-in-principle of inscribing tokens of symbols of a certain formal language), while Field assigns it to certain *spatio-temporal* objects (admitting as concrete entities regions of space-time that are irregular, disconnected, and of heterogeneous material content). Their books cast considerable doubt on the thesis of the scientific indispensability of mathematical objects.

Does this suffice to establish nominalism? Chihara and Field seem to think so. While for many readers the most valuable parts of Chihara's book will be the chapters on Russell and Poincaré, for the author himself, to judge by his Introduction, what is most important is the attempt to refute the anti-nominalist arguments of Quine, and some not dissimilar arguments of Kurt Gödel. Chihara implicitly presumes that a refutation of these arguments is tantamount to a proof of nominalism.

As for Field, his book bears the subtitle "A Defense of Nominalism," but includes (p. 4) the disclaimer that "nothing in this monograph purports to be a positive argument for nominalism." The resolution of the paradox lies in Field's presumption that nominalism does not need to be defended by positive arguments. He explicitly says that if he can accomplish the negative aim of undercutting the arguments of Quine and Putnam, then he will have reduced belief in mathematical objects to the status of "unjustifiable dogma." Thus Field, like Chihara, presumes the burden of proof to be on his lotus-eating "Platonist" opponent.

I disagree. Chihara and Field may have gone a long way toward showing that science *could* be done without numbers. I maintain, however, that science at present *is* done with numbers, and that there is no scientific reason why in future science *should* be done without them. And thus it is not the (continued) acceptance of mathematical objects, but rather the nominalist's insistence on their rejection, that constitutes an unjustified and anti-scientific philosophical dogmatism.

Quine and Putnam have been false friends of numbers in making the case for their acceptance seem to depend on a claim of indispensability. Actually, the burden of proof is on such enemies of numbers as Chihara and Field, to show either: (a) that science, properly interpreted, *already does* dispense with mathematical objects, or (b) that there are scientific reasons why current scientific theories *should be* replaced by alternatives dispensing with mathematical objects. I will call the claim (a) about the proper interpretation of current science *hermeneutic* nominalism, and the proposal (b) to replace current science by an alternative *revolutionary* nominalism.

I have argued that any anti-instrumentalist nominalism must be either hermeneutic or revolutionary. I will argue that hermeneutic nominalism, judged by the standards of linguistics, is an implausible hypothesis thus far unsupported by evidence; and that revolutionary nominalism, judged by the standards of physics, is a costly proposal thus far without scientific motivation.

3 HERMENEUTIC NOMINALISM

If we take everyday beliefs at face value, then we must conclude that natural numbers are posits of common sense dating from prehistoric times. If we take physics even halfway literally, then we must conclude that science has been committed to complex numbers for well over a century. According to hermeneutic nominalism, this is all illusion. General relativity theory may seem to make statements about vector-valued functions. Quantum mechanics may seem to make statements about linear operators. But, in fact, no physical theory asserts or presupposes the existence of such mathematical objects; no branch of science actually posits or commits itself to the existence of abstract entities.

Hermeneutic nominalism is thus a thesis of a type that has recently been described by Saul Kripke:

> The philosopher advocates a view in patent contradiction to common sense. Rather than repudiating common sense, he asserts that the conflict comes from a philosophical misinterpretation of common language – sometimes he adds that the misinterpretation is encouraged by the "superficial form" or ordinary speech. He offers his own analysis of the relevant common assertions, one that shows that they do not really say what they seem to say. (Kripke 1976, p. 269)

Let us imagine a laboratory assistant to Lord Kelvin reporting the data in some experiments on the conversion of mechanical to thermal energy. It sounds as if he is speaking of energy-in-joules and temperature-in-degrees-Kelvin and other such numerical and abstract entities. According to hermeneutic nominalism, he is actually speaking of something completely different: perhaps of possible chalk marks on possible blackboards (following Chihara). Maybe of so-called basic regions scattered through the vastness of space-time (following Field). Or perhaps of something still less expected and still more surprising (following some yet unwritten rival to Chihara (1973) and Field (1980)).

Now this claim is in itself not very plausible, and it becomes even less so when we reflect that to take anything like what we find in Chihara's book or Field's as an account of what the laboratory technician is saying is to attribute to that technician a tacit knowledge of such topics in foundations of mathematics as predicative analysis and measurement theory. These subjects did not even exist in Lord Kelvin's day, and even now they are studied by few pure mathematicians, let alone working physical scientists and their technical assistants.

Kripke's words (not specifically directed against nominalism by their author) seem appropriate here:

> Personally I think such philosophical claims are almost always suspect. What the claimant calls a "misleading philosophical misconstrual" of the ordinary statement is probably the natural and correct understanding. The real misconstrual comes when the claimant continues, "All the ordinary man really means is . . ." and gives a sophisticated analysis compatible with his philosophy.

Certainly the burden of proof is on the proponents of hermeneutic nominalism, who claim to have discovered a radical difference between appearance and reality in scientific discourse.

As a thesis about the language of science, hermeneutic nominalism is, I presume, subject to evaluation by the science of language, linguistics. For I am prepared to dismiss those who

> write as if, in addition to . . . everyday or "garden variety" rules of English, capable of being discovered by responsible linguistic investigation carried on by trained students of language, there were also . . . "rules" capable of being discovered only by *philosophers.* (Putnam 1971, p. 5)

In the current technical jargon of linguistics, the hermeneutic nominalist's thesis that scientific statements do not really say what they appear to say becomes the hypothesis that their *deep structure* differs from their *surface structure*, while the thesis that such statements are not really about what they appear to be about becomes the hypothesis that certain *noun phrases* in the surface structure are without counterpart in the deep structure.

Now readers of professional linguistics journals will recognize that hypotheses of this general type (though normally less radical than those of hermeneutic nominalism) are not seldom entertained by trained students of language. Such readers will also be familiar with the kinds of evidence cited in responsible linguistic investigations to support such hypotheses. Until some evidence of this kind can be adduced in support of its implausible hypotheses, I for one will be prepared to dismiss hermeneuticism as a desperate device of "ostrich nominalism."

4 REVOLUTIONARY NOMINALISM

It is one thing to observe that matters could equally well have been arranged otherwise than they currently are. It is quite another thing to urge that a rearrangement would constitute an *improvement.* To say that the British convention of driving on the left-hand side of the road is no worse than our own convention of driving on the right-hand side is not to advance a *criticism* of our traffic laws.

"Science," Putnam tells us, lives "extremely happily on the rich diet of impredicative sets" (1971, p. 56). The work of Chihara and Field suggests that science could survive on more meager fare, on a diet of inscription-possibilities or of spatiotemporal regions. But would science be *healthier* after such a change of menu?

When scientists abandoned caloric fluid and luminiferous ether, it was because they had discovered alternative theories that were *empirically* superior, of wider scope and greater accuracy in predicting the results of observations and experiments. Now the alternative theories concocted by Chihara and Field cannot be claimed to be empirically *superior* to current theories, for they have been designed to be empirically *equivalent.*

Will it be urged that those alternatives are somehow *pragmatically* superior? Their awkward and ungainly character makes it difficult to claim that they are more convenient and efficient as systematizations of the data of experience. Will it be urged that, despite their unnatural and artificial character, they somehow contributed to clarity, simplicity, intelligibility, and the like, in ways that matter to working scientists? Something of the sort *must* be urged if a nominalistic revolution in science is to be *motivated.*

The proviso, "in ways that matter to working scientists," is crucial, if a mere instrumentalist opposition to science is to be avoided. It is pointless and futile to urge a revolution in the practice of *physicists* motivated only by considerations appealing only to *philosophers* of a certain type. Physicists are too well aware of the dismal historical record of philosophical interference in science to accept such dictation from outsiders.

Now the avoidance of ontological commitments to abstract entities does not seem to have won recognition in the scientific community as being in itself a goal of the scientific enterprise on a par with scope and accuracy, and convenience and efficiency, in the prediction and control of experience. It seems, on the contrary, a matter to which most working scientists attach no importance whatsoever. It seems distinctively and exclusively a preoccupation of philosophers of a certain type. Thus Goodman is able to cite only a few linguists who are nominalistically inclined, and not one physicist:

> Paucity of means often conduces to clarity and progress in science as well as philosophy. Some scientists indeed – for example, certain works in structural linguistics – have even imposed the full restriction of nominalism upon themselves in order to avoid confusion and self-deception.

One would search the *physics* journals in vain for any expression of nominalistic qualms and scruples, of reluctance and hesitancy to use mathematical apparatus, of suspicion that such "Platonistic claptrappery" as complex numbers may be a source of "confusion and self-deception."

The proposed nominalistic revolution in physics can be scientifically motivated only by showing that the avoidance of ontological commitments to abstract entities would somehow serve indirectly to advance us toward some more recognizably scientific goals. For my own part, I cannot discern any such scientific *benefits* to be expected from the proposed revolution, while I do discern a couple of non-negligible *costs*.

First, any major revolution involves transition costs: the rewriting of textbooks, redesign of programs of instruction, and so forth. A reform along the lines of Chihara (1973) would involve reworking the mathematics curriculum for science and engineering students, avoiding impredicative methods in favor of predicative parodies that are harder to learn and not so easy to apply.

A reform along the lines of Field (1980) would involve reworking the physics curriculum, so that each basic theory would initially be presented in qualitative rather than quantitative form. A course on measurement theory would have to be crammed into the already crowded study plan, to explain and justify the use of the usual numerical apparatus. This is educational reform in precisely the wrong direction: away from applications, toward entanglement in logical subtleties.

Second, the physicist who puts on nominalistic blinders may be unable to see certain potentially important paths for the development of science. I have in mind here not an inevitable logical consequence of nominalistic revolution, but a likely psychological consequence. Chihara (1973, p. 209) promises that he will recant his nominalism should some future physical theory turn out to require mathematical objects indispensably. But the danger I have in mind is that if science goes nominalist today, that future theory *may simply never be discovered.* Yuri Manin has noted this point in connection with intuitionism:

> Unfortunately, it seems that it is these "extremes" – bold extrapolations, abstractions which are infinite and do not lend themselves to a constructivist interpretation – which make classical mathematics effective. One should try to imagine how much help mathematics could have provided twentieth century quantum physics if for the past hundred years it had been developed using only abstractions from "constructive objects." Most likely, the standard calculations with infinite dimensional representations of Lie groups which today play an important role in understanding the microworld, would simply never have occurred to anyone. (Manin 1977, pp. 172–3)

(Mention of quantum mechanics should remind us that it is unclear whether the methods of Chihara and Field are adequate even for present-day science in its entirety. For Chihara the problem is a minor one, and could probably be solved by adopting a somewhat stronger system of predicative analysis than the particular weak system Σ_ω he favors. For Field, the problem is a major one, for he has given us no idea how he proposes to treat quantum theory, which differs radically (owing to its use of *infinite-dimensional* apparatus and to its *statistical* character) from the one theory he does treat in detail, Newtonian gravitational theory.)

But I need not enlarge on the costs for present-day and future physics of a nominalistic revolution. Surely the burden of proof is on the revolutionary, who proposes a drastic departure from our thus far eminently successful policy of ontological tolerance in common sense and scientific theory construction. Until it is shown that nominalism offers physical science some substantive advantages, I for one am prepared to dismiss its revolutionary proposals as motivated only by medieval superstition ("Ockham's razor") and fastidious bigotry (cf. Goodman 1964, objection viii).

Chihara and Field have gone a long way toward constructing nominalistic alternatives empirically equivalent and pragmatically only slightly inferior to our current scientific theories. Their work suggests that an ontology of abstracta may be one feature of those current theories that is merely *conventional*, in the best sense of the word (that of David Lewis 1969). This suffices to cast considerable doubt on some more *extreme* versions of realism.

It does not suffice to cast doubt on *moderate* versions of realism, which merely observe that our current theories seem to invoke abstracta and that we do not yet have reasons to abandon those theories. For to characterize some feature of our present ways of doing things (in scientific theorizing or in driving) as conventional is not in itself to criticize that feature. And Chihara and Field have not come close to constructing nominalistic alternatives that are manifestly *superior* (empirically or pragmatically) to our current scientific theories.

5 NOMINALISM, ONTOLOGICAL AND EPISTEMOLOGICAL

I have rejected nominalism in its traditional *ontological* form, as the doctrine that there exist no abstract entities. I equally reject it in its currently fashionable *epistemological* form, as the thesis that even if there exist any abstract entities, still we could never come to know about their existence. Epistemological nominalism is usually supported by an argument of the following form: all entities of which we can have knowledge are causally connected with our organism; no abstract entities are causally connected with our organism; ergo, no abstract entities are entities of which we can have knowledge.

The argument is, of course, valid, a syllogism in Camestres. But the premises are dubious and debatable. As for the minor premise, of course a cyclic group does not act on our organs of sight, touch, and hearing in the same way as an alarm clock. And nobody nowadays, except perhaps a stray numerologist or two, would imagine that mathematical objects act on us through some mysterious sixth sense or ESP unknown to orthodox physiology. Nonetheless, as Maddy (1980) skillfully argues, there is a good deal of research in developmental psychology and neurophysiology that can be read as showing that we do, in a sense, have causal contact with certain abstracta.

As for the major premise, it rests on a *causal theory of knowledge*. That theory has many opponents, who regard it as a half-truth arrived at by over-hasty generalization from too narrow a range of cases, to which the cases of knowledge of mathematical objects, ethical values, other minds, and so forth are just so many counterexamples. Significantly, that theory has also a good many half-hearted sympathizers, who do not regard it as wrong-headed or misguided, but merely as in need of amendment. In many amended versions, the notion of causality disappears, to be replaced by that of reliability or explanation or something of the sort, and with it disappears the major premise of the epistemological nominalist's syllogism. Again Maddy (1984) provides a useful survey of the issues.

The more cautious sympathizers with the causal approach to the theory of knowledge now maintain only that the abstractness and consequent causal inertness and isolation of mathematical objects create difficulties for the epistemologist in trying to account for mathematical knowledge. I am surprised to find Field citing these epistemological difficulties as if they in themselves constituted some sort of grounds for nominalism:

> [Nominalism] saves us from having to believe in a large realm of . . . entities which are very unlike the other entities we believe in (due for instance to their causal isolation from us and from everything that we experience) and which give rise to substantial philosophical perplexities because of those differences. (Field 1980, p. 98)

A footnote to this passage makes it plain that Field's "philosophical perplexities" are precisely the epistemological difficulties just alluded to. (Incidentally, the same footnote provides a good bibliography of works arguing for epistemological nominalism.)

To bring out just how odd this argument is, I want to consider a parallel: suppose that Burrhus Skinner were to confess that after all those years of work with his rats and pigeons he is still "substantially perplexed" by the ability of freshman students to master calculus and mechanics. Now what mathematician or physicist would take that as motivation for rewriting the textbooks in those subjects? What linguist would take it as evidence that the sentences in those textbooks have some bizarre and outré depth grammar?

No one would take it as an indication of anything but the inadequacies of behaviorist learning theory.

Likewise, a philosopher's confession that knowledge in pure and applied mathematics perplexes him constitutes no sort of argument for nominalism, but merely an indication that the philosopher's approach to cognition is, like Skinner's, inadequate.

CONCLUSION

Unless he is content to lapse into a mere instrumentalist of "as if" philosophy of science, the philosopher who wishes to argue for nominalism faces a dilemma: he must search either for evidence for an implausible hypothesis in linguistics, or else for motivation for a costly revolution in physics. Neither horn seems very promising, and that is why I am not a nominalist.

APPENDIX

For the reader's convenience, I here outline the constructions of Chihara and Field, and the claims which those authors make for their constructions. I will not advance any *technical* objections against those constructions (though in fact I have one small reservation about Chihara's approach, and share with Kripke several large reservations about Field's), since my aim has been to argue that even if the constructions are technically flawless, they do not suffice to establish nominalism.

A Chihara's modal nominalism

I here outline the construction of Chihara (1973), Chapter V and Appendix. Chihara's strategy is to reinterpret in a nominalistically acceptable fashion a portion of pure mathematics: arithmetic first, then so-called *predicative* analysis. He then argues that the portion of mathematics so reinterpreted suffices for scientific applications, and dismisses the rest of mathematics (the impredicative part) as mythology.

To illustrate Chihara's approach to arithmetic, consider Euclid's famous theorem:

(0) $(\forall$ number $m)(\exists$ number $n)(m < n$ & n is prime)

As a first attempt to avoid mathematical objects, let us rewrite this as:

(1) $(\forall$ numeral $a)(\exists$ numeral $b)\ldots$

(I will indicate only the transformation of the *prefix* of (0); this is not to say that the transformation of the *matrix* does not require some caution.) Now if numerals are taken as *types* (patterns of inscription), then they are themselves abstract entities akin to shapes, and (1) is not much of an improvement on (0). But if numerals are taken as *tokens* (individual inscriptions), then they are concrete entities, made of chalk or ink, but there may not be (indeed, almost certainly are not) enough of them around to make (1) true. To get a version of (0) that is both true and committed only to concrete entities, we must introduce the *modal* notions of necessity (□) and possibility (◇). Then, taking numerals as tokens, our final reinterpretation of (0) is:

(2) □(∀ numeral *a*)◇(∃ numeral *b*)...

Informally this says: however long a tally you could ever write down, I could write down a still longer one such that ... Here we have the idea behind the approach to arithmetic in Chihara (1973).

Chihara's approach builds on the work of *predicativists* (specifically, Hao Wang), mathematical constructivists somewhat more liberal than intuitionists. Predicativists accept uncritically classical *arithmetic* (theory of natural, or equivalently rational, numbers) but in *analysis* (theory of real numbers, or equivalently of *sets* of natural numbers) they accept only what is definable. To begin with, they accept those sets of natural numbers that are definable by purely arithmetical conditions, conditions quantifying only over natural numbers. These are the *order zero* sets. Next they accept those sets of natural numbers that are definable by conditions quantifying over natural numbers and order zero sets. These are the *order one* sets. And so on, through higher and higher orders. (Just how high to go is a delicate question.) A surprisingly large portion of classical mathematics can be "parodied" within this framework, as the survey by Feferman (1977) shows. Intuitively, it is plausible that a theory of definable sets should be reducible to arithmetic plus truth-predicates, with quantification over definable sets being replaced by quantification over the code numbers of their defining conditions, and the membership relation replaced by the relation "*n* is the code number of a formula with one free variable that is *true* of *m*." The details can be worked out, and we get a reduction of predicative analysis to something that has already been shown to be nominalistically reinterpretable. Chihara's account of the application of mathematics in science is illustrated by Figure 2.1.

While scientific theories are formulated mathematically in terms of sharply defined functions, at least in the overwhelming majority of

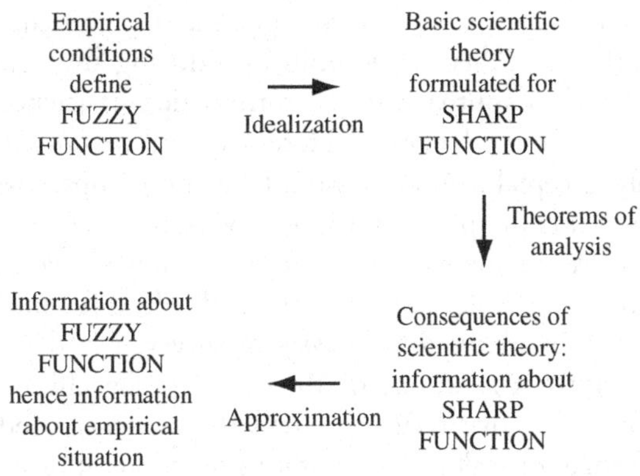

Figure 2.1

applications empirical conditions define only fuzzy functions. For instance, the condition:

$f(t) = x$ iff
the projectile is x meters above the floor of the chamber
at t seconds after firing

cannot define a sharp function because of the fuzziness of the projectile and the chamber (viewed on a scale of micrometers) and of the firing event (viewed on a scale of nanoseconds). Thus the application of a scientific theory to empirical conditions typically involves an element of idealization. Now predicative mathematics provides sufficiently many sharp functions to serve as idealizations for empirically defined fuzzy functions, because any classical function can be approximated as closely as desired by a predicative function. Moreover, the theorems of analysis used in deriving consequences from basic scientific theories can all be parodied predicatively: this can be verified by comparing the mathematics curriculum for science and engineering students as listed in any college catalogue with the survey by Feferman (1977) of predicative mathematics. Thus predicative mathematics, which we have already seen to be nominalistically reinterpretable, suffices for scientific applications.

B *Field's spatiotemporal nominalism*

I here outline the construction of Field (1980). Field's strategy is to reformulate basic scientific theories and their consequences (the special information they entail about special situations) in a way that avoids all mathematical vocabulary, and then to argue that the consequences can be deduced from the basic theories without introducing any mathematics.

To illustrate Field's approach to the formulation of science, let us consider, as he does, thermodynamics. Here a typical *qualitative*, math-free, nominalistically acceptable notion would be the comparative relation R between point-events of space-time given by "x is cooler than y." Here a typical *quantitative*, mathematical, "Platonistic" notion would be that of temperature on a given scale, conceived as a real-valued function r on point-events in space-time. *Measurement theory*, as surveyed in the compendium Krantz *et al.* (1971), is a corpus of theorems to the effect that suitable assumptions on qualitative relations entail the existence (and uniqueness up to stated changes of scale) of quantitative functions appropriately representing them. In thermodynamics, suitable assumptions would include that R is irreflexive and transitive, appropriate representation would include that xRy if and only if $r(x) < r(y)$, and stated changes of scale would be like those used in passing between Fahrenheit and Celsius. Once we have such a basic *representation theorem*, it becomes possible to reformulate any scale-invariant assumption on the quantitative functions as an assumption on the qualitative relations. In thermodynamics, *continuity* for the temperature function r can be reformulated in terms of a notion of *temperature-basic region*, itself defined in terms of the cooler relation R.

In this way the nominalist can reformulate the whole of science, both basic, general theoretical principles, and particular consequences for practical applications. However, the only route we have seen so far from the qualitatively formulated version of a basic theory to the qualitatively formulated

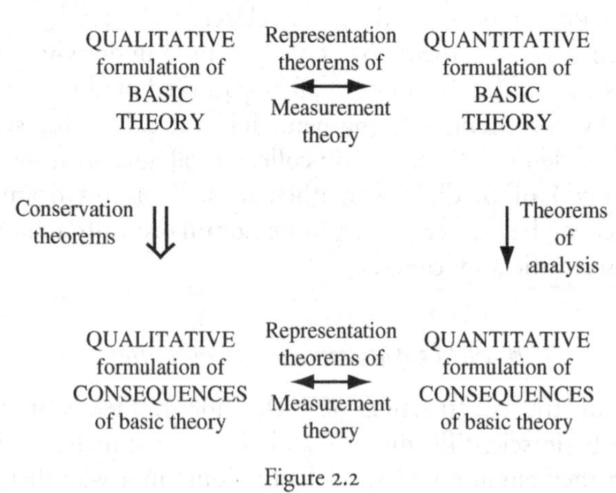

Figure 2.2

versions of its consequences involves a "Platonistic" detour: from qualitative basic theory by measurement theory to quantitative basic theory, thence by theorems of analysis to quantitative consequences, and thence by measurement theory again to qualitative consequences, as in Figure 2.2.

It is, however, theoretically possible, though practically inconvenient, to avoid the introduction of mathematics, to avoid the detour through the quantitative and abstract:

> the conclusions we arrive at by these means are not genuinely new, they are already derivable in a more long-winded fashion . . . without recourse to the mathematical entities. (Field 1980, pp. 10–11)

> for these purposes ["problem solving"] the usual numerical apparatus is a practical necessity. But it is a necessity that the nominalist has no need to forgo: he can treat the apparatus . . . as a useful instrument for making deductions from the nominalistic system that is ultimately of interest; an instrument which yields no conclusions not obtainable without it, but which yields them more easily. (Field 1980, p. 91)

These claims are supported by appeal to *conservation theorems* from *proof theory* (the most important being perhaps one due to Scott Weinstein).

3

Mathematics and Bleak House

1 "NOMINALISM" AND "REALISM"[1]

Nominalism is a large subject. In our book (Burgess and Rosen 1997) my colleague Gideon Rosen and I distinguished a negative or destructive side of nominalism, which tells us not to believe what mathematics appears to say, from a positive or reconstructive side, which aims to give us something else to believe instead. We noted that there were a few nominalists who contented themselves with the negative side, conceding that mathematics is useful, insisting that what it appears to say is not true, and letting it go at that, without attempting any reconstrual or reconstruction of mathematics. We expressed some surprise that there were not more such destructive nominalists, since as compared with reconstructive nominalism, destructive nominalism has what Russell in another context called "the advantages of theft over honest toil"; and if nothing else was clear from the work of Hartry Field, Charles Chihara, Geoffrey Hellman, and other reconstructive nominalists whose work we surveyed, it was clear that the amount of honest toil that would be required for a nominalistic reconstrual or reconstruction of mathematics would be quite considerable.

Today, a couple of years after publication, it is beginning to seem that the main achievement of our book will have been to provide a decent burial for the hard-working, laborious variety of nominalism. For almost everything that has come forth since from the nominalist camp has represented a light-fingered, larcenous variety, which helps itself to the utility of mathematics, while refusing to pay the price either of acknowledging that what mathematics appears to say is true, or of providing any reconstrual or reconstruction that would make it true. The usual label for this variety of

[1] I have decided to keep this paper in the form in which it was originally written for oral delivery, adding footnotes to supply citations of the literature, and for clarification at a few points where experience has shown misunderstanding of my intended sense may be likely.

nominalism is "[mathematical] fictionalism."[2] Not only has fictionalism become the most widely pursued form of nominalism, but even some former anti-nominalists have been wavering or drifting in its direction, including even the author of one of the most eloquent early criticisms, "Mathematics and *Oliver Twist*." I will take the liberty, therefore, of substituting "fictionalism" for the "nominalism" label in the official title of our symposium.

While I am at it, I had better say something about the "realism" as well. It is an even larger subject, since there is hardly any bit of philosophical terminology more diversely used and overused and misused than the R-word. There seems to be a systematic difference between the way in which the word is understood by many of those who describe themselves as "realists" and the way in which it is understood by most of those who describe themselves as "anti-realists." For many professed "realists," realism amounts to little more than a willingness to repeat in one's philosophical moments what one says in one's scientific moments, not taking it back, explaining it away, or otherwise apologizing for it: what we say in our scientific moments is *all right*, though no claim is made that it is *uniquely* right, or that other intelligent beings who conceptualized the world differently from us would necessarily be getting something wrong. For many professed "anti-realists," realism seems rather to amount to a claim that what one says to oneself in scientific moments when one tries to understand the universe corresponds to Ultimate Metaphysical Reality; that it is, so to speak, a repetition of just what God was saying to Himself when He was creating the universe.

The weaker position might be called anti-anti-realism, and the stronger position capital-R Realism. Quine says somewhere that his "realism" and his "pragmatism" are reconciled by his "naturalism." This is a hard saying, but I think it can be explained along the following lines. First, "realism" here means anti-anti-realism: the refusal to apologize while doing philosophy for what is said while doing mathematics or science. Second, "pragmatism"

[2] In earlier publications I called this stance "instrumentalism," while Rosen called it "constructive nominalism," by analogy with van Fraassen's "constructive empiricism." (Incidentally, though van Fraassen is best known for his fictionalism about unobservable physical entities, he is also a fictionalist about abstract mathematical entities.) However, the analogous positions in other areas of philosophy (e.g. modality) are today generally called "fictionalist," and on this ground Rosen adopted the "fictionalist" label, and I now follow him. Unfortunately, there has been another use of "fictionalist" in philosophy of mathematics, by Hartry Field and his students, for whom it includes all *rejectionist* views, all views that hold that standard mathematics is false, whether fictionalist in our sense or reconstructive. This usage seems to be out of alignment with the usage of "fictionalism" in other areas of philosophy, and for that reason to be avoided.

means rejection of capital-R Realism, rejection as unjustified of any claim that mathematics and science give us a God's-eye view of capital-R Reality. Third, "naturalism" means adherence to a conception of epistemology as an inquiry conducted by citizens of the scientific community examining science from the inside, rather than an inquisition conducted by philosophers foreign to science judging science from the outside.

And the reconciliation? Naturalism teaches us to look at our scientific, philosophical, and other forms of intellectual endeavor as activities of biological organisms with cognitive capacities that, though extensive, stop well short of omniscience. As such, *none* of these endeavors can succeed in achieving a God's-eye view of Reality. And therefore there is no reason to apologize for *one* of them, science, failing to achieve such a view, and every reason *not* to suppose that another of them, philosophy, could do better.

Should I, therefore, replace "realism" by "naturalism" in the title of our symposium? No, for unfortunately "naturalism" too has been used and overused and misused in multiple senses. In particular, while in Quine's sense naturalism abstains from imposing philosophical constraints on science, there is another equally widespread but diametrically opposed conception in which "naturalism" consists precisely in imposing such a philosophical constraint, namely, the constraint that no entities are to be assumed that do not stand in natural cause-and-effect relations with us. So I will just replace "realism" by "anti-fictionalism," making the title of our symposium come down to this: "fictionalism in philosophy of mathematics: pro and con."

2 LITERARY GENRES

To begin with the pro side, it is impossible to quarrel with the proposition that mathematics is *in some respects* like fiction. For indeed, *anything* is like *anything* else, *in some respect.* I even think the comparison may be illuminating, as to the nature of fiction, or rather, as to some of the questions philosophers raise about fiction, and in particular about the status of fictional characters. The hard-headed view is that they just do not exist. Another view is that they are abstract entities of some sort or other. Yet another view is that they are *mental* entities. Now the view that mathematical entities are mental entities, though a perennial favorite of amateur philosophers, has been in disrepute among professional philosophers since Frege's trenchant critique of psychologistic philosophies of mathematics more than a hundred years ago. I think that fiction is enough like mathematics to suggest that the view that fictional entities are mental entities is equally dubious. For example, one problem for mentalistic theories of mathematical entities is that there are *too many* such entities for minds to have created each one; and a little reflection shows that the situation is similar with fiction: a novelist may write of the doings of a vast army with thousands of officers and myriads of soldiers, while in fact describing only a couple of dozen of them individually. But I am straying from our topic, which was supposed to be what, if anything, comparison with fiction can show us about mathematics, and not the reverse.

Reverting to that topic, then, I have said on the one hand that it can hardly be denied that mathematics is like fiction in some respects, most obviously in consisting of large bodies of manuscript and printed and now electronic writing. On the other hand, it must be said that there is a sense in which mathematics is clearly non-fiction: there is a well-established practice of classifying writing as "fiction" and "non-fiction," and setting aside attempted theoretical definitions and analyses, attending rather to the criteria by which in actual practice such classifications are made, clearly mathematics counts as non-fiction. The compilers of the *New York Times* best-seller list will never put any mathematical work, however wonderful, at the top of the fiction column, and not just because nothing even by Andrew Wiles will ever sell like Stephen King. Nor will any librarian catalogue, say, the *Proceedings of the Cabal Seminar* as an "anthology of short stories based on the characters created by Georg Cantor." Now of course misclassifications are sometimes made, and mischievous hoaxes and pious frauds sometimes succeed, but these represent mistakes in applying general criteria to specific cases. I do not think it even makes sense to suggest that the general criteria are *themselves* mistaken, and it seems unquestionable that by those criteria mathematical writing is not literally fictional writing: mathematics is not *in all respects* the same as fiction.

So the question is: in which respects is mathematics like, and in which respects is it unlike, fiction? That in part depends on the species of the genus fiction one considers. The first species one thinks of will probably be the novel, and comparison with novels is in fact common among professed fictionalists and their critics, as with *Oliver Twist*, mentioned earlier. There has been, however, a minority – including especially the late Leslie Tharp[3] – who have preferred a comparison with mythology. In a slightly different vein, Steve Yablo, in an interesting paper,[4] has suggested a comparison with metaphor. Metaphor of course is not a genre of fiction but a figure of

[3] Rescued from oblivion in Chihara (1989).

[4] The reference was to an on-line version, but the paper has since appeared in print as Yablo (2000).

speech, and I think that to speak as Yablo does of a metaphor running on for volumes and volumes and volumes is to stretch the concept of "metaphor" well beyond breaking point. But perhaps much of the content of Yablo's suggestion could be preserved if we took the comparison to be between mathematics and parables or fables.

Be that as it may, I believe the comparison with fables is the most apt of the candidates I have considered, and comparison with novels the least so. Novels almost always are attributable to identifiable individual authors: Proust or Flaubert, Trollope or Dickens. Some fables are attributable to such authors, Lafontaine for instance; others are traditional. Mathematics also consists of both traditional elements and elements with identifiable authors. Novels are almost always unique. Fables tend to be retold over and over in variant versions by different writers, so that we have Aesop's version, Lafontaine's version, and many latter-day retellings of the fox and the crow, for instance. Mathematics likewise gets retold by textbook writer after textbook writer. The characters in one novel seldom reappear in another, and even those who do reappear, like Swann or Palliser, do so only in comparatively few stories, all by the same author. This is so with some characters of fable, but many, like the clever fox, reappear in whole cycles of tales. The same mathematicalia, π and e, the sine and cosine functions, 0 and 1 and 2 and so on, reappear through whole libraries of mathematical works. Again, characters encountered in novels are generally of the same species as those encountered in daily life, while those in fables are, as one dictionary definition reminds us, beings of a different order: "animals that talk and behave like human beings." Mathematics, too, has objects even more unlike those of any other subject, and it is for precisely that reason that there is thought to be a philosophical problem about them.

Yet more important is the matter of application, which in literature typically takes the form of a "message." The fable typically though not invariably has a "moral," while to demand one of the novel is virtually the definition of Philistinism. I am reminded in this connection of what Nabokov says in his posthumously published lectures about *Bleak House* and its supposed concern with reform of the Court of Chancery.

> At first blush it might seem that *Bleak House* is a satire. Let us see. If a satire is of little aesthetic value, it does not attain its object, however worthy that object may be. On the other hand, if a satire is permeated by artistic genius, then its object is of little importance and vanishes with its times while the dazzling satire remains, for all time, as a work of art. So why speak of satire at all? . . . Such cases as Jarndyce did occur now and then in the middle of the last century although, as legal historians have shown, the bulk of our author's information on legal matters

goes back to the 1820s and 1830s so that many of his targets had ceased to exist by the time *Bleak House* was written. But if the target is gone, let us enjoy the carved beauty of the weapon. (Nabokov 1980, p. 64)

The question of applications is crucial in the case of mathematics, because though it would be a kind of Philistinism to demand that every piece of mathematics have one, many do; and it is precisely *because* many do that many philosophers have opposed nominalism, this being the least common denominator of all "indispensability arguments."

Still more important, however, is a feature common to all genres of fiction. The most important single respect in which fictionalists hold mathematics to be like novels or fables or whatever is in being a body of *falsehoods*. In particular the existence theorems of mathematics are supposed to be untrue: these say there exist, for instance, prime numbers greater than $10^{10^{10}}$, whereas according to mathematical fictionalists, and indeed all nominalists, there are no such things as numbers at all.[5]

3 NON-LITERAL LANGUAGE

Nominalism, to repeat, is a large subject. In our book, Rosen and I distinguished two varieties of reconstructive nominalists. Those of one variety, the *hermeneutic*, insist that their reconstruals of mathematics reveal what, contrary to superficial appearances, deep down mathematical language has meant all along: mathematical theorems are true, but while what mathematical theorems appear to mean implies the existence of mathematical entities, what they really mean does not. This position might be summed up in the formula, "There are no numbers, and some of them are primes greater than $10^{10^{10}}$." Reconstructive nominalists of the other variety, the *revolutionary*, concede that their reconstructions of mathematics are not analyses of current mathematics, but amendments to it; not exegeses, but emendations.[6]

[5] I have been concerned here with comparison between mathematics and fiction less for its own sake than for its bearing on the issue of nominalism. The comparison deserves a much fuller examination. For an extended discussion of analogies and disanalogies, containing extensive references to the further literature, see Thomas (2000, 2002).

[6] This is not the place to discuss reviews of Burgess and Rosen (1997) at any length, but it may be mentioned that many adherents of and sympathizers with reconstructive nominalism have wished to claim that there is some third alternative. For instance, it is sometimes said that a nominalist interpretation represents "the best way to make sense of" what mathematicians say. I see in this formulation not a third alternative, but simply an equivocation, between "the empirical hypothesis about what mathematicians mean that best agrees with the evidence" (hermeneutic) and "the construction that could be put on mathematicians' words that would best reconcile them with certain philosophical principles or prejudices" (revolutionary).

We did not in the book discuss a division between hermeneuticists and revolutionaries among the fictionalists, but such a distinction can be made.[7] The hermeneutic fictionalist maintains that *the mathematicians' own understanding* of their talk of mathematical entities is that it is a form of fiction, or akin to fiction: mathematics is like novels, fables, and so on in being a body of falsehoods *not intended to be taken for true*. According to the hermeneutic fictionalist, the anti-nominalist philosopher is being more royalist than the queen (of the sciences); is being a kind of fundamentalist, *taking literally what was never so meant*.

Consider, for instance, the question when the "reification" of numbers first occurred historically, in the main line of development leading up to modern mathematics. What I take to be a fairly conventional view dates the reification of natural numbers to about the time of Archytas of Tarentum and other Pythagoreans, among whom we begin to see the transition from the use of numerals as adjectives, as in "six oxen were sacrificed to Zeus," to their use as nouns, as in "six is a perfect number"; and it dates the reification of real numbers from about the time of Omar Khayyam and other medieval Islamic and Hindu mathematicians, who in contrast to Greek mathematicians treated ratios of geometric magnitudes as numbers, as things that could be added and multiplied. According to Chihara (1990), however, these dates are all wrong. The reification of number actually occurred not circa 500 BCE or 1000 CE, among mathematicians, but some time in the 1960s and 1970s, and among philosophers: it is a rash and recent innovation of Quine and other "literalist" philosophers, myself and my fellow symposiast included.

Chihara's argument for this claim is essentially that the mathematicians *he* has questioned have either expressed puzzlement when asked about their ontological commitments, or have repudiated suggestions that they are committed to any ontology at all. Yablo's argument is on the face of it different, but I believe at bottom similar. He begins by calling attention to a very interesting phenomenon, the "paradox" of his title. Let me attempt my own presentation of it.

According to an old story, when Lindemann settled the ancient problem of squaring the circle, his colleague Kronecker reduced him to tears by

[7] The term "hermeneutic fictionalism" was taken as the title for a large-scale study by Jason Stanley (2001), which pursues at length a number of the points to be made below, and many more also, through several different areas of philosophy where fictionalism has become fashionable. In an alternative terminology, hermeneutic reconstructive nominalism and hermeneutic non-reconstructive nominalism are called *content* hermeneuticism and *attitude* hermeneuticism respectively. This terminology is used in Rosen and Burgess (2005).

asking, "What is the value of your investigation of π, since irrational numbers do not exist?" Suppose a nominalist philosopher today were to say to Wiles, "What's all this fuss about your proving there are no natural numbers x, y, $z > 0$ and $n > 2$ such that $x^n + y^n = z^n$? Since there are no numbers at all, a fortiori there are none satisfying that equation."[8] I imagine most mathematicians would be contemptuous of this speech and most philosophers – even most *nominalist* philosophers – embarrassed by it.

According to another story, G. E. Moore famously once argued along something like the following lines: he held up his hands and said, "As you can see, here are two human hands. Since human hands are material bodies in the external world, there exists an external world of material bodies." Suppose an anti-nominalist philosopher today were to hold up his hands and say, "As you can see, the number of my hands is two. Since the number two is an abstract entity, abstract entities exist." Would not most mathematicians be baffled and bewildered by this argument, and would not most philosophers – even *anti-nominalist* philosophers – balk and boggle at it?

On the one hand, objections to what mathematicians and others would ordinarily say about mathematical entities on the grounds of their alleged non-existence are regarded with scorn, while on the other hand, purported proofs of their existence are viewed with suspicion. Yablo argues that this double attitude is explicable if we assume that mathematical assertions are meant non-literally, as fiction, or rather, in this preferred terminology, as metaphor. For that would make both objections on the grounds that they are not literally true, and attempts to prove that they *are* literally true, equally off the mark in opposite directions. As I said earlier, I think Yablo's argument ultimately relies on the same kind of consideration as Chihara's, namely, mathematicians' puzzlement at or repudiation of philosophical theses and arguments about the existence of the entities they study. I also think the positions of Chihara and Yablo ultimately involve similar mistakes about the nature and meaning of "commitment" and "literalness."

Let me begin with commitment. The reason Quine and others have spoken of the "commitment" of mathematical, scientific, and everyday thought to mathematical and other abstract entities, is precisely because they did *not* want to speak of mathematicians' or scientists' or lay-persons' "assertions" of or "beliefs" in the existence of such entities. Mathematicians *qua* mathematicians do address questions about whether there are prime numbers greater than $10^{10^{10}}$, but they generally do *not* spend much time talking, and presumably do not spend much time thinking, about the

[8] This example is from a talk by the late George Boolos.

question whether there are any such things as numbers at all: hence the inappropriateness of speaking of their "assertions" or "beliefs" about such questions. Quine's claim was that they are *committed* to an affirmative answer to this question, because what they *do* assert and believe, that there are prime numbers greater than $10^{10^{10}}$, *implies* that there are prime numbers and therefore that there are numbers, and it has this implication whether or not they ever acknowledge it, and indeed even if they repudiate it, when talking philosophy rather than mathematics.

Now for literalness. Quine does allow that someone might *say*, "There are prime numbers greater than $10^{10^{10}}$" and yet *not* be committed to its implications. For one could say it and *not really mean it*, or *not really believe it*, whether or not one were capable of articulating what it is that one *does* mean and believe when saying it. One could say it and not intend it to be understood "literally." But what does this mean? If the function of the word "literal" were to indicate the *presence of something positive*, some extra enthusiasm perhaps, then it would indeed be very doubtful whether mathematicians who assert that there are prime numbers greater than $10^{10^{10}}$ mean it "literally." But I suggest that the function of the word is in actual fact less to indicate the presence of something positive than to indicate the *absence of something negative*: roughly what in the legal phrase is called "mental reservation and purpose of evasion." If this is so, then what is in actual fact very doubtful is whether mathematicians who assert that there are prime numbers greater than $10^{10^{10}}$ intend their assertion only as something "non-literal." To do *that*, they would have to be more philosophically self-conscious than they appear to be. To put the matter another way, the "literal" interpretation is not just one interpretation among others. It is the *default* interpretation. There is a *presumption* that people mean and believe what they say. It is, to be sure, a *defeasible* presumption, but some *evidence* is needed to defeat it. The *burden of proof* is on those who would suggest that people intend what they say only as a good yarn, to produce some actual evidence that this is indeed their intention.[9]

[9] To mean what one says literally is simply to mean what one says, just as to be a genuine antique is simply to be an antique. The force of "literally" is not to assert that one is doing something more besides, but to deny that one is doing something else instead: meaning something *other* than what one says, as when one speaks metaphorically, hyperbolically, elliptically, or otherwise figuratively. One does not have to think anything *extra* in order to speak literally: one has to think something extra in order to speak *non*-literally. Such, at any rate, is the not uncommon view of the meaning of "literal" to which I subscribe. For more on this point, see Searle (1979). Assuming this point, the hypothesis that someone is writing or speaking literally is the hypothesis that nothing more is going on than meets the eye or the ear: the *null* hypothesis about covert intentions. In the text I call it the *default* hypothesis, since I take it there is a defeasible presumption in favor of null hypotheses in general. In the specific

And I submit that the fact that mathematicians tend to be perplexed by and dubious about philosophical argumentation over the existence of mathematical entities – the "frog and mouse battle" as Einstein once called a specific instance of such debate – and even the fact that mathematicians who are badgered with skeptical questions by skeptical philosophers can be got to make skeptical noises, are not very good evidence. In this connection, I remember reading an interview in *Scientific American* some years back with Murray Gell-Man, who explained that in his original paper on the quark hypothesis he avoided claiming that quarks were real in order to avoid trouble with "philosophers." This self-censorship perhaps does not say much for Gell-Man's courage – Galileo, after all, did not recant until he was "shown the instruments" – but it is rather revealing as to how we philosophers are viewed by some of our scientific colleagues, and suggests that there may be serious difficulties with the methodology of pestering scientists for opinions on philosophical issues to which they may have given little or no thought, and accepting their answers as indicative of their intentions in putting forward the affirmations that they do put forward when philosophers leave the scene and let them get back to work.

4 CONTRASTING CASES

To underscore these points, I would like to contrast the comparative lack of evidence for attributing a *global* "non-literalist" intention to mathematicians *generally* with two cases where we *do* have good evidence of "non-literalist" intent. One case is that of certain *local* usages and idioms, the other of certain *specific* mathematicians.

To take the latter case first, among professional mathematicians there are a tiny minority who have given serious, sustained thought to philosophical questions. Notoriously they tend to disagree with each other quite as much as professional philosophers do – think of Hilbert and Brouwer – so that the quickest way to tell whether they are talking mathematics or talking philosophy is to listen whether they are agreeing or disagreeing. Probably among this very small and sharply divided group of mathematician–philosophers can be found adherents of virtually every position found

case of the null hypothesis about *meaning*, there is the additional consideration, not mentioned in the text, that word-meaning and speaker-meaning, though distinct, are not independent. It would be impossible for words to mean what they do if everyone always used them to mean something else, and difficult for them to mean what they do unless *most* people *most* of the time use them to mean that, so that a randomly chosen person at a randomly chosen time *probably* means what he or she says.

among philosophers proper, including positions that regard all of mathematics as just a good yarn, or merely a great game.

Certainly the bulk of Hilbert's own results, beginning with his basis theorem, were meaningless or "ideal" statements according to his official philosophy. The topological work, and notably the fixed-point theorem, which established Brouwer's reputation among mathematicians, is largely false or meaningless according to his intuitionist principles. There is of course always a question, whenever someone says something in one context and takes it back in another, whether we should regard the original affirmation as merely pretense and the subsequent denial as revealing the person's real opinion, or whether contrariwise we should regard the ceremony of recantation as play-acting, and the real opinion the one expressed originally. But I think that in the case of the field marshal of the batrachians and the generalissimo of the rodents we may conclude on the basis of their extensive philosophical writings that they did indeed mean much of what they said in their mathematical work as "non-literal" or "fictitious" in some sense. My earlier point was that in the case of the overwhelming majority of mathematicians there is no such evidence that unspoken philosophical caveats accompany their mathematical assertions.

The other case I wish to mention is that of the mathematician who, for instance, asserts on one page that there is only *one* non-cyclic group of order four, and on the next page that among the subgroups of some larger group there are *three* non-cyclic groups of order four. How can there be three of them if there is only one of them? That is a mystery whose solution is that the assertion of uniqueness was not meant literally, but rather involved a figure of speech, namely *ellipsis*: "unique" was elliptical for "unique up to isomorphism," and what is really meant is that all non-cyclic groups of order four, including the three that turn up as subgroups of the larger group alluded to, are isomorphic to each other. This isomorphism is, of course, what the *proof* of "uniqueness" proves.

The difference between such cases of mathematicians *locally* meaning this or that form of expression as one or another kind of figure of speech, and the hermeneutic fictionalist's claim that mathematicians globally mean *none* of what they say literally, is that there *is* evidence *within what mathematicians say while engaged in mathematical research or teaching*, to indicate, for instance, that the claim of uniqueness of the Klein four-group is intended to be understood as pertaining not literally to uniqueness, but to uniqueness up to isomorphism. For one thing, if a student who had not yet learned all the relevant idioms and usages were to raise a question, the mathematician would explain. Owning up to the non-literal character of the uniqueness assertion is something the mathematician does *on the job*, not just when interrupted and pestered by skeptical philosophers, or after hours pursuing some philosophical hobby.

To be sure, it is also possible in this particular example that the mathematician just does not know what "unique" means, like the merchant whose advertisement I once saw, who claimed that "Every item in the store is unique, and many are one-of-a-kind." But there are other classes of examples of *local* non-literalness. Indeed, my discussion of the case of uniqueness assertions is largely inspired by Stewart Shapiro's discussion of one extensive class of such cases,[10] mathematicians' frequent use of dynamic language, as if the functions, for instance, were *moving* mathematical objects around, *changing* one mathematical object into another, and so on – a manner of speaking to which, as Shapiro notes, Plato already objected. I think again, with Shapiro, that there is good evidence *within what mathematicians say while engaged in mathematical research or teaching* for taking this particular kind of language to be meant non-literally. But the presence of evidence in this case again contrasts with the comparative absence of evidence for the hermeneutic fictionalist's global claim that all kinds of mathematical language are meant "non-literally."

5 UNDECIDABLE QUESTIONS

My conclusion, then, is that hermeneutic fictionalism is implausible, and that if one is going to be a fictionalist, one had better be a revolutionary fictionalist, denying while doing philosophy what is asserted while doing mathematics, but not pretending that it never *was* asserted, or pretending that it was only asserted but was not really *meant* or *believed*. And there are indeed similarities between mathematics and fiction that can be cited by the revolutionaries that would seem to give no support at all to hermeneutics. One such is the apparent *incompleteness* of mathematics, suggest by the Gödel theorems. (I trust that in speaking to an audience of logicians I need hardly add that the question of the philosophical bearing of these theorems is by no means an easy one, and that such theorems are, in the words of my fellow symposiast, the "beginning of the story" and not "the end of the story.") This is often compared to the incompleteness of fictional tales, which leave many questions, beginning with the proverbial length of the protagonist's nose, undecided and undecidable.

[10] In public talks since the later 1970s and in Shapiro (1997).

This similarity would seem to provide less than no support to the hermeneutic view that *mathematicians do* intend their mathematical assertions as fictions. After all, mathematicians were making mathematical assertions for centuries before Gödel, in blissful ignorance of any incompleteness or incompletability phenomena, and even while confidently asserting that "in mathematics there is no *ignorabimus*" and "*wir müssen wissen, wir werden wissen.*" So the fact that fiction rather obviously leaves many questions undecided would seem to be a powerful argument for the conclusion that mathematicians were *not* thinking of their science as a branch of creative writing. By contrast, the observation that mathematics resembles fiction in respect of incompleteness might well be thought to provide support for the revolutionary view that *philosophers should* regard mathematical assertions as fictions. At any rate, the observation often *is* cited in support of such a position.

To this sort of observation there are two kinds of response: a more specific and a more general. The more specific notes that, even if we accept that Gödelian theorems do show that there are fundamental reasons of mathematical principle why some mathematical questions must remain forever undecided, this by no means establishes a dividing line with mathematics and novels, fables, and so on, on the one side, and scientific and commonsense thought on the other side. For are there not, to begin with, also fundamental reasons of physical principle why some physical questions must remain forever undecided? The second law of thermodynamics suggests the growth of entropy will blur the historical record to the point where many facts about the past will become irrecoverable. General relativity suggests information may be swallowed up by black holes and hidden from us forever beyond their horizon. And then there is quantum mechanics. On top of all this, there are the apparently undecidable questions that arise wherever there is vagueness, which is just about everywhere *except* mathematics. But these considerations are doubtless familiar, and I need not enlarge on them here.

What I want to consider instead is another line of response, suggesting that *even if it were* distinctive of mathematics among what pass for sciences to present us with unanswerable questions, it would be doubtful that this feature should be taken as a criterion for distinguishing fiction from fact. More generally, the other line of response I wish to consider – one vaguely analogous to Hume on miracles – suggests that it is virtually always more doubtful that philosophy has arrived at the correct understanding of "truth" or "existence" or what have you, than that what well-established mathematical, scientific, and commonsense principles tell us are facts are fantasies, or that what they tell us exists is a phantasm.

Revolutionary mathematical fictionalism is an "error theory," offering a "correction" to mathematics, and especially to mathematical existence theorems, much as Brouwer once offered an intuitionistic "correction" to his own Brouwer Fixed Point Theorem.[11] At the same time, it is offering a "correction" to science insofar as it is formulated mathematically. The line of response against such philosophical "corrections" to mathematics and science that I want to consider is the one expressed in the *Credo* of my colleague David Lewis. Rosen and I quoted it in our book, and I will not quote it again here; but for those who have not read it, its point is that, given the comparative historical records of success and failure of philosophy on the one hand, and of mathematics on the other, to propose philosophical "corrections" to mathematics is *comically immodest.*

And indeed, the historical record of philosophical "corrections" to mathematics and science from Bellarmine's "correction" of Galileo – an early form of fictionalism or "constructive empiricism" – onwards, has been pretty dismal. One argument against revolutionary fictionalism is thus just that, given the historical record, on simple inductive grounds it seems extremely unlikely that philosophy can do better than mathematics in determining what mathematical entities exist, or what mathematical theorems are true, and much more likely that for the $(n+1)^{\text{st}}$ time, philosophy has got the nature of truth and existence wrong.

6 INTERMINABLE DEBATE

Supplementing this Ludovician line of thought, there is another, deriving ultimately from Rudolf Carnap's "Empiricism, semantics, and ontology" (Carnap 1950), that deserves attention. It is a line of thought admittedly in considerable disrepute today. For instance, the late George Boolos in a passage in Boolos (1997b) characterizes the Carnapian argument I have in mind as "rubbish." And so far as I know, when that paper was delivered orally this remark raised no substantial murmur from the audience. Any attempt to rehabilitate an argument that has been thus consigned to the rubbish bin of philosophical history by so prominent a figure and with so little protest would be a difficult undertaking, and what I will attempt here will be something much less ambitious than a full rehabilitation. I will merely attempt to restate the essence of the argument in a way that strips it of most of its dated formulations and presuppositions.

[11] The theorem can be found in Brouwer (1976) and the "correction" in Brouwer (1975).

To begin with, Carnap notes, as so many others have since, and as I did above, that the question whether there exist any such things as numbers at all is one that is never raised by the mathematicians themselves (while doing mathematics), but only by philosophers (including such few professional mathematicians as are also amateur philosophers, when doing philosophy rather than mathematics). And the most salient fact Carnap notes about the debate among philosophers is that it goes on and on and on without ever being settled. The philosophical issue of nominalism and realism is in this respect like *Jarndyce and Jarndyce*.

There is an apparent difference in that in Dickens' novel the parties to the lawsuit were finally driven mad by the interminable legal wrangling, whereas few philosophers have been driven over the edge by the issue of nominalism and realism. But then philosophers are not really in the position of the suitors. Rather, as Rosen put it when we were once discussing the matter, "We're the lawyers." Of course, in Dickens' novel, even the *lawyers* eventually stopped arguing the case: the case ended when the entire value of the estate had been eaten up by legal costs, and there was no money left to *pay* any more lawyers. So far we philosophers are still being paid, and even to some slight degree paid attention, and perhaps there is no immediate danger of the value of the estate running out in our case. Still, Carnap thought that the endless litigation with no sign of settlement approaching was an indication that something was badly wrong.

His analysis of the situation was rather complex, and had three strands: there is a positive phase with two aspects, plus a negative phase. Let me take the elements one at a time. The two aspects of the positive phase can be brought out by comparing Carnap's position on the one hand with that of one of Yablo's targets, Crispin Wright, and on the other hand with that of all nominalists' target, Quine.

Wright once propounded an argument against nominalism roughly as follows: I have as many fingers as toes; but as everyone who understands the concept of "number" knows, to say that I have as many fingers as toes is equivalent to saying that the number of my fingers equals the number of my toes; but to say this presupposes that there is such a thing as the number of my fingers or toes; hence the number ten exists. What the Carnapian agrees with in this argument is the recognition that concepts come with rules for their employment, some of which entail affirmative answers to certain existence questions, so that one has only two choices: either one rejects the concept, in which case the existence questions cannot even be asked; or else one accepts the concept, in which case one immediately gets affirmative answers to those existence questions. One cannot ask the question and answer it in the negative. One is compelled to accept certain existence assertions, if one accepts the concept, or the "framework" to substitute Carnap's term for Wright's.

One is not, however, compelled to accept every concept that might be proposed. In response to criticism by Hartry Field, Wright acknowledged this point, and in subsequent rounds of debate he and his colleague Bob Hale have attempted to formulate a general principle that *would* enforce acceptance of the concept involved in many though not all cases, including the particular case relevant to Wright's original argument, that of the concept of "number."[12] So far they have not come up with a general principle that has commanded any very widespread support, and the Carnapian view would be that it is a mistake to attempt to enforce the acceptance of concepts by a priori arguments.

Quine, and along with him Hilary Putnam at one stage in the evolution of his views, urged a very different sort of reason for accepting the existence of numbers (or other abstract mathematical entities to which numbers could be "reduced"). According to Quine, we must (alas, with the greatest reluctance!) resign ourselves (ah, that it should have come to this!) to accepting (unbidden and unwelcome!) mathematical entities, because (most regrettably and unfortunately!) mention of them seems (would that it were not so!) to be an unavoidable requirement (how cruel a necessity!) in formulating scientific theories. What Carnap agrees with in this argument is that the ultimate grounds for accepting the concept "number" is the role it plays in formulating scientific theories, together with the role scientific theories in turn play in much of our life.

Quine's argument invites two kinds of objections. On the one hand, there are the reconstructive nominalists who question that mention of mathematical entities really is an unavoidable necessity after all. It is the large concessions to nominalism made in the anti-nominalist argument of the ex-nominalist Quine that invite this sort of objection. On the Carnapian view, by contrast, a "framework" of mathematical entities is not something to which we grudgingly resign ourselves because it is a necessary evil, but something we gratefully adopt because it is an enormous convenience; and there is no suggestion that if mathematical entities by hook or crook or somehow *could* be eliminated, then they *should*, and hence no invitation to nominalists to go looking for hooks and crooks.

[12] See Hale and Wright (2001), especially "Responses to Critics."

On the other hand, there are objections of the kind especially voiced by Charles Parsons (1980, p. 150), roughly to the effect that Quine's argument does not do justice to the seeming *obviousness* of elementary arithmetic, and makes acceptance of "two plus two is four" or "the number of my hands is two" depend on recondite and abstruse considerations about whether it is possible to formulate general relativity without referring to tensors, or quantum mechanics without reference to linear operators. Such objections are invited by Quine's refusal to acknowledge any "conceptual" truths, even *conditional* ones, and his insistence that all knowledge "faces the tribunal of experience" as a *whole.* By contrast, on the Carnapian view, the seemingly obvious simple statements about the number two just mentioned are taken to be immediate, *given* the relevant concept or framework, and considerations of the larger role of that framework in science are only relevant to explaining why the framework, as well as whatever comes thus immediately with it, is classified as non-fiction rather than fiction. Moreover, to repeat, what is important about the larger role in science is merely that it would be *very inconvenient in practice* not to use mathematical language, and not that detailed scrutiny of the most sophisticated theories shows it to be *wholly impossible in principle* to do without such language.

It is against the background of this two-sided positive account, according to which there are both a local a priori element, the rules constitutive of the concept, and also a global pragmatic element, motivating acceptance of the concept, that Carnap attempts to account for the interminability of philosophical debate between nominalists and anti-nominalists. What is missing, according to Carnap, when philosophers debate whether numbers exist – not whether they exist *according to standard mathematics*, or whether it is *convenient to speak and think as if* they existed, but whether they *really* exist – is any framework providing rules for assessing assertions of ontological metaphysics about what italics-added *really* or capital-R Really exists, in the way that the framework of number-language determines that "I have as many hands as feet" is a sufficient condition for "The number of my hands equals the number of my feet." Or to put the matter another way, the ontologists are treating the practical question whether to accept the framework of number-language, to which only considerations of convenience and the like are genuinely germane, as if it were a theoretical question of a kind that can only be asked within a preexisting conceptual framework. Or to put the matter yet another way, and as briefly as possible, the debate never ends because there is not only a lack of agreement between the two sides as to what the right answer to the question is, but also a lack of agreement between the two sides, and for that matter even among those on the same side, about *what counts as a relevant consideration for or against* one or another answer to the question.[13]

7 MEANINGLESS QUESTIONS

I suspect the reason Carnap's presentation of the case failed to convince was largely that he was too much identified with the infamous "empiricist criterion of meaningfulness," which certainly has by now been consigned, if not to the rubbish bin, then at least to archives, where it may be studied by *historians* of philosophy, but where it no longer influences *current* philosophical debate. According to this criterion – which among other things sits rather poorly with a recognition of Gödelian phenomena – the absence of agreed empirical and/or logical criteria for what counts for or against one or another ontological hypothesis renders argumentation for these theses *meaningless*, a kind of nonsense poetry without the poetry. Instead of a comparison of mathematics with the work of Charles Dickens or George Elliot, we have here a comparison of philosophy of mathematics with the work of Lewis Carroll or Edward Lear: it makes the same amount of sense, though it is less entertaining. Carnap was notorious for the provocative claim that the issue of nominalism and realism, the "problem of universals," is a "pseudo-problem." This Carnapian thesis is much stronger than, for instance, the Ludovician claim that induction suggests it is less likely that philosophy has now solved the problem of universals in a way that shows mathematics to be in error, than that philosophy has once again failed to solve the problem. Moreover, the Carnapian thesis implies or presupposes other very large and controversial claims.

Let me elaborate. One very traditional sort of way to try to make sense of the question of the ultimate metaphysical existence of numbers would be to turn the ontological question into a theological question: did it or did it not happen, on one of the days of creation, that God said, "Let there be numbers!" and there were numbers, and God saw the numbers, that they were good? According to Dummett, and according to Nietzsche – or my

[13] As mentioned in the review (Burgess 2001), there is a superficial appearance of similarity between Carnap's view and that of Mark Balaguer, especially as in Balaguer (1998, pp. 158–79), who maintains that the question of the existence of numbers is "factually empty." Really, there is a deep difference between the two positions, since Balaguer does not distinguish two questions as Carnap does. Like someone who thought that the plain, literal meaning of "Prime numbers greater than $10^{10^{10}}$ exist" was "Prime numbers greater than $10^{10^{10}}$ are ingredients of ultimate metaphysical *Reality*," Balaguer declines to join Carnap in answering, "Yes, of course," to the question, "Do numbers exist?" If doubters as well as deniers of numbers are counted as nominalists, then his refusal to return an affirmative answer to the question whether numbers exist makes Balaguer a kind of nominalist.

perspective on Nietzsche – this is the *only* way to make sense of questions of ontological metaphysics. The Carnapian claim that ontological metaphysics is meaningless is roughly equivalent to the conjunction of this Nietzsche–Dummett thesis, "realism makes sense only on a theistic basis," with *analytic atheism*, the thesis that theological language is meaningless. Both these theses are highly controversial: analytic atheism was explicitly rejected even by outspoken agnostics like Russell, and the Nietzsche–Dummett thesis is rejected by many philosophers in Australia who regard themselves as simultaneously "realists" in some strong sense, and "physicalists" in some sense equally strong.

I myself believe, like Russell, that analytic atheism is false, and suspect, contrary to the Australians, that the Nietzsche–Dummett thesis is true. If as I believe the theological question does make sense, and if as I suspect it is the only sensible question about the italics-added *real* or capital-R Real existence of numbers, then I would answer that question in the negative; but then I would equally answer in the negative the question of the *Real* existence of just about anything. For as has been said elsewhere, everything we have learned about our processes of cognition points in the direction of the conclusion that even other intelligent creatures, to say nothing of an Omniscient Creator, would or might well have patterns of language and thought very different from ours, recognizing categories of objects very different from those we recognize, or perhaps not even having a category of "objects" at all, if they used a language without a category of nouns, as well they might.

Since I do not wish to claim that the absence of empirical meaning is tantamount to the absence of all meaning, where Carnap would put forward a categorical negative, "These questions are meaningless," I only put forward a rhetorical question, "*What* are these questions supposed to mean?" But I do agree with Carnap that the question of the *Real* existence of mathematical entities does lack *empirical* meaning, and while I do not think this settles the question of nominalism, I think it does have an important bearing on the question of how much mathematics is like novels, fables, and other forms of fiction.

For consider the question whether, say, the works of Carlos Castañeda or Rigoberta Menchù are non-fiction or fiction: are they eye-witness reportage of magic or tragic occurrences, or merely novels masquerading as anthropology or autobiography? In these cases, we know very well what it would have *looked* like for the events in Menchù's book to have occurred, and in this age of cinematic special effects, we can even say the same for Castañeda's books. By contrast, as regards the question of the ultimate metaphysical *Reality* of numbers, we have absolutely no idea of what difference it would make to how things *look*; or rather, we have a very strong suspicion that it would make no difference at all. This is what is meant by having an *empirically* meaningful question in one case, and an *empirically* meaningless question in the other.[14]

I think that in view of this radical difference between mathematics and novels, fables, or other literary genres, the slogan "mathematics is a fiction" is not very appropriate, and the comparison of mathematics to fiction not very apt. My conclusion is that, whatever may remain to be said for or against nominalism, about whether we should or should not call ourselves "nominalists," we should not call ourselves "fictionalists."

[14] In other words, I mean no more by saying that the choice between two views is "empirically meaningless" than that the two views themselves are empirically equivalent. Thus understood, it is a trivial truism that there is no empirically meaningful difference between any given theory T, and the fictionalist alternative T^* according to which everything observable happens as if theory T were true, though it is not. What would be highly non-trivial would be the claim that the difference between T and T^* is meaningless *tout court et sans phrase*. But that would only follow from empirical meaninglessness assuming a discredited empiricist criterion of meaningfulness.

4

Quine, analyticity, and philosophy of mathematics

I TWO SENSES OF "FOUNDATIONS OF MATHEMATICS"

Does mathematics requires a foundation?[1] The first thing that must be said about the question is that the expression "foundations of mathematics" is ambiguous. Let me explain.

Modern mathematicians inherited from antiquity an ideal of rigor, according to which each mathematical theorem should be deduced from previously admitted results, and ultimately from an explicit list of postulates. It also inherited a further ideal according to which the postulates should be self-evidently true. During the great creative period of early modern mathematics, there were and probably had to be many departures from both ideals. But during the century before last, as mathematicians were driven or drawn to consider less familiar mathematical structures, from hyperbolic spaces to hypercomplex numbers, the need for rigor was increasingly felt, and higher standards were eventually instituted. But while the ideal of rigor may be claimed to have been realized, the ideal of self-evidence was not.

Considering only the ideal of rigor, the working mathematician's understanding of its requirements, of what is permissible in the way of modes of definition and modes of deduction of new mathematical notions and results from old, is largely implicit. Logic, which investigates such matters, and fixes explicit canons, is a subject in which the algebraist, analyst, or geometer need never take a formal course. Nor are mathematicians in practice much concerned with tracing back the chain of definitions and deductions beyond the recent literature in their fields to the ultimate

[1] This question was the title of the Arché Institute conference, August 12–15, 2002, where a preliminary version of this paper was first delivered. I would like to thank the leadership of the institute and local organizers of the affair, and especially Crispin Wright and Fraser MacBride, for making my participation both possible and pleasurable, and my fellow participants for valuable feedback on the preliminary draft.

primitives and postulates. But there is a place, and one even may go so far as to say a need, for *someone* to investigate the choice of postulates, and what differences a different choice would make. It is these kinds of investigations, when carried out by mathematical means, that in standard classifications of the branches of mathematics are called "logic and foundations." Or rather, that label is applied to the kind of investigation just mentioned plus any other research that can fruitfully apply the same methods.

This, then, is the first and weaker of two senses in which the term "foundations" is used. There will be a need for "foundations of mathematics" in this first sense so long as mathematicians continue to adhere to an ideal of rigor, and – despite hype from some popularizers about "the death of proof" – that would mean for the foreseeable future.

But there is a second and stronger sense, in which one would speak of a "foundation" for mathematics only if, in addition to the ideal of rigor, the ideal of self-evidence or something like it were realized. Though I have listed this sense second, it is presumably older, since it is only if something like the ideal of self-evidence is realized that the metaphor implicit in the word "foundations" is appropriate. Postulates with something like self-evidence would provide the firm foundations of the edifice of mathematics, and this firmness together with the firmness of the rigor by which new results are built upon old would guarantee the firmness of the higher stories. This picture is merely the application to mathematics of the picture offered by epistemological "foundationalism," according to which the edifice of knowledge is to be built up from a secure and privileged basis by secure and privileged means.

Mathematicians have learned to live with the absence of self-evident postulates, of "foundations," and in this sense passively acquiesce in the proposition that foundations (in the foundationalist sense) are not required. Many philosophers remain more troubled by the situation, and in consequence either lapse into nominalist, constructivist, or other heresies, rejecting orthodox mathematics, or involve themselves in programs to provide the missing foundations.

For a very familiar specific instance of the distinction between two senses of "foundations," we may consider the result we have been taught to call *Frege's theorem*, namely, the deducibility of the basic laws of arithmetic from the postulate we have been taught to call *Hume's principle*, according to which, if there are as many of *these* as of *those*, then the *number* of the former is equal to the *number* of the latter. Uncontroversially Frege's theorem is a major contribution to foundations of mathematics in the first and weaker sense, which is concerned with logical relationships between postulates and theorems, without too much concern over the status of the postulates. But there has been a controversy, involving the late George Boolos and the nominalists on one side, and some of the St Andrews school on the other, over the status of Hume's principle, and over whether Frege's theorem can provide a foundation for mathematics in the second and stronger sense.

On one view, Hume's principle is analytic, Frege's theorem does provide such a foundation for arithmetic, and the challenge is to find a way of providing a similar foundation for more of mathematics. On the other view, Frege's theorem does *not* provide a foundation, and Hume's principle is either a synthetic truth, or an untruth. What I want to do here is to elaborate a third or intermediate position, according to which Hume's principle *is* analytic, but still does *not* provide a foundation for arithmetic in the sense of foundationalist epistemology. Naturally this presupposes a notion of analyticity in which a statement may be analytic but nonetheless need not be self-evident or a logical consequence of self-evident statements, or anything of the sort. In sketching the intermediate position I will be mainly concerned to sketch a conception of analyticity with this feature.

My starting point will be, as my title suggests, the thought of the late W. V. Quine. His work, by the way, provides another illustration of the distinction between the two senses of foundations. Quine was, in his generation, a significant contributor to "logic and foundations" in the first sense. (I heard it said, by one of my fellow speakers at a memorial meeting, that asked about his standing on one occasion, he described himself as "captain of the B-team"; and this seems quite a just estimate.) But Quine was also famously a paradigmatic opponent of epistemological foundationalism, and the author of the best-known rival to the architectural metaphor. According to Quine, knowledge is not a building but a web, more or less fixed at the edges by the attachment of observation sentences to sensory evidence, but underdetermined as to how we spiders should spin the middle portions, including mathematics, which lies somewhere very near the center.

My reexamination of Quine will be a sequel to an earlier reexamination of Carnap, entitled "Mathematics and *Bleak House*" (Burgess, 2004b). I even thought of entitling the present paper "Mathematics and *Bleak House* II," though in the end I was deterred from doing so when I found myself having only one occasion to refer to the Dickens novel, mentioning the police investigator in it, one Bucket, who was for a generation[2] the

[2] Until the appearance of Sherlock Holmes.

canonical fictional detective in the English-speaking world. The concern of my earlier paper with the Dickensian title was to re-evaluate Carnap's classic "Empiricism, semantics, and ontology" (Carnap 1950). Here I wish to consider Quine's reply, "Carnap's views on ontology" (Quine 1951a) and more particularly the famous paper of a few months earlier on which that reply was based, "Two dogmas of empiricism" (Quine, 1951b).

To put the matter very roughly, Quine argued in replying to Carnap that the position of Carnap presupposed the analytic-synthetic distinction, the first of the two dogmas Quine took himself to have refuted. Like some other recent commentators,[3] I dissent from the common view that Quine clearly vanquished Carnap in their exchange. To put matters very roughly again, my claim will be that Quine almost *needs* to recognize a notion of analyticity – and also that he *can* recognize such a notion, without betraying his core philosophical principles.

2 QUINIANISM VS PLATONISM

Before I examine the differences between Quine and Carnap, I wish to consider what divides them both from the nominalists. And before I consider what divides them from the nominalists, I wish to consider what Quine, Carnap, and many nominalists have in common that divides them from anyone who would really deserve to be called a "Platonist" in anything like a traditional sense. I will begin with a sympathetic description of – I do not pretend it is anything like an *argument* for – a Quinian world-view.

It was the ambition of Galileo, Kepler, and other worthies of their era, by close reading of the book of nature, to discover the very intentions of its great Author in writing it; or, to vary the metaphor, to produce a plan of the universe faithfully reproducing the blueprint used by its great Architect in constructing it. For Quine, this is a hopeless ambition: no science produced by human beings can provide a God's-eye view of the universe, and this should be evident almost as soon as one begins to view the human knower as an object of scientific study.

When human knowers are so viewed, human knowledge, including especially scientific theorizing, is seen as the product of certain organisms

[3] Let me in particular cite two useful unpublished works, the Princeton senior thesis of Tom Dixon "Separating Semantics from Empiricism and Ontology," and the doctoral dissertation of Inga Nayding, "Positing Existence." Both see less difference between Quine and Carnap than perhaps the two saw between themselves, and both attempt, each in his or her own way, to narrow the difference still further. I derived encouragement from their example, even though my own way of attempting to narrow the difference is not quite theirs.

in a certain environment, seeking to fulfill certain needs. Beavers build dams; people first construct hydrological theories, and only *then* construct dams (unless, having also constructed ecological theories, they decide *not* to construct the dams after all). Scientific theories make intelligible patterns in the environment as we experience it, and in favorable cases make it possible to influence (or warn us not to try to influence) the course of that experience. But what science can make intelligible in our experience, and how it can make it so, inevitably depends on the nature of our intellects, and what kinds of experience we are capable of. Thus scientific theory, product of a certain organism in a certain environment, will inevitably be the way it is in part because that environment, the universe, is the way it is, and in part because the organisms, ourselves, are the way we are: there is no reason to suppose intelligent extraterrestrials, with very different kinds of sensory experience and very different intellects, would produce the same scientific theories we have.

For that matter, there is not much reason to believe that even if we ourselves had it to do all over again we would come up with the very same theories we have. Thus scientific theories are the way they are partly because the universe is the way it is, partly because we are the way we are, and partly because of a third factor: partly because the interaction between the universe and ourselves has gone the way it has. But if our theories as they are thus differ from what they might equally well have been had history gone slightly differently, and differ even more from what the theories of alien creatures might be expected to be like, then a fortiori they must differ greatly from the "theories" of an omniscient Creator, and the ambition of gaining a God's-eye view of the universe must be unrealizable. Such is the Quinean picture at the highest level of generality.

To descend to a level of slightly greater specificity, one feature of the way our intellects work is that language is crucial to scientific thought, and our language exhibits a comparatively limited range of grammatical forms. In particular, our scientific theories run very much to sentences of the noun–verb, subject–predicate, object–property type. As we employ sentences of this type over and over in different contexts and with different functions within scientific theorizing, our scientific theories will be positing over and over again different kinds of objects with different kinds of properties.

To be a little more specific still, all this applies the mathematical apparatus deployed in our scientific theories. Starting with the use of numerals as adjectives, we have found that to bring out certain patterns a shift to using them as nouns is required, and so natural numbers have been posited. After long speaking only of relations of proportions among geometric magnitudes, we have found it immensely convenient, if not practically indispensable, to shift to speaking of ratios of magnitudes as objects, and as objects that can be added and multiplied, and so real numbers have been posited. Later we have found useful a transition from speaking of real numbers (plural) with certain properties to speaking of the set (singular) of such real numbers, thus positing sets as single objects constituted by pluralities of objects.

We thus end up speaking of different kinds of numbers, sets, functions, and so forth in sentences of the same grammatical form as ones about medium-sized dry goods, even though these sentences occupy very different positions and roles in the body of our knowledge. "Septimus is a prig" and "Seventeen is a prime," for instance, have similar grammatical or logical forms, but very different epistemological positions. The best way to verify the former would be to locate Septimus in space and time and interact with him causally: he may sit near us and speak, sending soundwaves to our ears, by which we detect his priggishness. The number seventeen does not do anything analogous. It does not sit in Cantor's paradise and shine N-rays on our pineal glands, by which we detect its primality. That arithmetical property is checked by quite different means.

This feature of our actual scientific theories is perhaps one they need not have had. Whether or not we could have managed to do without any non-spatial, non-temporal, causally inactive, causally impassive mathematical objects at all – the partial success of programs of nominalistic reconstruction of mathematics suggests that in principle this might have been possible, though examination of the details suggests that in practice it might not have been feasible – there is no strong reason to suppose that if we had it to do all over again we would end up with the very same *kinds* of mathematical objects. As for the scientific theories of space aliens, not only is there no strong reason to suppose they would involve distinctive mathematical objects, but what is more, there is no strong reason to suppose they would involve objects at all, since there is no strong reason to suppose space aliens would have a language that involved nouns. The Chomskians maintain that universal grammar is *species specific*, and in combinatory and predicate-functor logics we get at least a vague and dim image of what a language with a grammar radically unlike ours might be like. And as for the "theories" of a Deity, "we see but through a glass, darkly."

Thus Quine – and in this Carnap would surely join him – can concede to the nominalist that the (particular kinds of) mathematical objects that figure in our current scientific theories are there largely because of the way we are (and the way our interaction with the universe has gone) rather than because of the way the universe is. The positing of numbers may be extremely convenient and in practice even indispensably necessary for us, but theories that involve such posits cannot be claimed to give a God's-eye view of the universe, to reflect ultimate nature of metaphysical reality, or anything of the sort.

3 QUINIANISM VS "FICTIONALISM"

What Quine – and again Carnap would surely be with him – will not concede to the nominalist is that any of this gives us any reason at all to reject current science and mathematics. It gives us no reason why we should perform, on entering the philosophy seminar room, a ceremony of abjuration, recanting what we habitually say outside that room, when engaged in scientific work or in daily life. *No* theory of *ours* can give a God's-eye view of the universe: "the trail of the human serpent" will be over all. But the fact that any particular theory fails to deliver a reflection of the ultimate nature of metaphysical reality, uncontaminated by any contribution from *us*, is no special grounds for complaint against that particular theory. If there are grounds for complaint, it is against the human condition.

And thus Quine is willing to speak inside the philosophy room in the same terms in which he speaks outside it, neither taking back, nor explaining away, nor apologizing for, the use of number-laden or set-laden language. Rather he is willing to reiterate in his capacity as philosopher the theorems of mathematics and theories of science. To apply the words used by the great Scottish philosopher in a somewhat different context, "Nothing else can be appealed to in the field, or in the senate. Nothing else ought ever to be heard of in the school, or in the closet."[4]

Here we have direct opposition to the most common kind of nominalist today. They *do* take back in their philosophical moments what they assert in their scientific moments. And for most of them, that is about *all* they do by way of expression of their nominalist allegiances: few nominalists any more are involved in active programs for reconstructing science on a number- and set-free basis. This kind of inactive nominalism is generally called "fictionalism"[5] and in my earlier paper on Carnap it was the contrast between his

[4] David Hume, "On the practical consequences of natural religion, *or*, Of a particular providence and a future State," in *Enquiry Concerning the Human Understanding*, Part XI.

[5] The label is also used by some activists like Hartry Field, whose views I am leaving to one side in the present discussion.

views and fictionalism that was my topic. Much that I said there about Carnapian anti-fictionalism applies also to Quinian anti-fictionalism, and I will recapitulate briefly.

The first thing to be said against "fictionalism" is that the label is remarkably ill chosen. To say, for instance, that Victorians regarded *Bleak House* as fiction, is among other things to say that when Victorians were in need of the services of a good detective, they did not waste any time looking for Bucket. They did not use the contents of the novel as a guide to practice.[6] The label "fictionalist" for the dominant contemporary variety of nominalism is ill chosen because those nominalists who apply the label to themselves do *not* in fact regard current mathematically formulated ordinary and scientific lore and theory in the same way they regard the productions of Dickens or other novelists. They *do* use such lore and such theory as a guide to practice.

The analogy "fictionalists" cite ought to be, not with works of imaginative literature, but with scientific theories that are regarded as no more than useful approximations to more complex but truer theories, known or remaining to be developed. An architect, for instance, designing a modest private residence, will obviously have to take into account the topography of the site, the fact that some points on the site are at a higher elevation above sea level than others, but generally will *not* take into account the curvature of the earth, or the fact that points at the same elevation do not lie on a perfect plane. To this extent, the architect is using as a guide to practice the primeval theory that the earth is flat, though no architect today believes any such thing.

Here is a case where a theory is rejected, a theory is not believed, and yet that theory is *not* regarded as fiction, as a work of creative writing, but rather is used as a guide to practice, is employed for practical purposes. The analogy the mislabeled "fictionalists" ought to be citing is with such cases, with uses of flat-earth geography when we know the earth is round, or of Newtonian gravitational theory when we know it is only an approximation to Einsteinian general relativity, or indeed of this last when we know it needs a quantum correction, though we do not as yet have a developed theory of quantum gravity.

But the citation of such examples soon makes evident a serious disanalogy between the attitude of the nominalist and that of the scientist or engineer who makes use of a theory known to be only a simplified approximation.

[6] At least not in the way they would have used it as a guide had they thought it non-fiction: for all I know, some readers may have been stirred by it to work for reforms in the Court of Chancery.

This can be brought out by considering an architectural firm suddenly given a commission for a much larger project than the private homes that are all they had built before. They would need to take into account the fact they had been ignoring, that the earth is round, and recalculate whether it is safe to ignore its curvature on the new and larger scale on which they would be working. If the project were as vast as the Very Large Array,[7] clearly they could not get away with treating the earth's surface as flat, they way they can in building a house on a half-acre lot. For projects of intermediate size, they would have to think at least for a moment about the question, or ask some consulting engineer whether flat-earth geography can still be trusted. And the engineer would base the calculation on round-earth geography: just how far flat-earth theory can be trusted for architectural purposes is something that is estimated in terms of round-earth theory.

The situation is similar in all cases of technical and everyday applications of theories that are not really accepted in the sense of being not really believed. The scientific and technical application of an approximation is always framed by some kind of estimate of how good the approximation is, obtained from some more accurate and credible theory, whether fully developed or still a work in progress.

The situation is quite dissimilar in the case of the nominalists' attitude towards mathematics, pure and applied. Here there is no question of the untruth of ordinary and scientific theories *ever* being relevant to practical questions. Nor is there any nominalistic alternative theory present in the background, or being sought. Rather, "fictionalism" became the dominant form of nominalism in the 1990s largely as a result of disappointment with the search for nominalist alternatives to standard mathematical formulations of scientific theories in the 1980s. The dissimilarity and disanalogy I have been describing marks nominalism as a *non-*, *un-*, or *anti-scientific*, and distinctively *philosophical* concern. For a Quinean, this feature – a willingness to say that some formulation is acceptable in everyday and *all* scientific contexts, however theoretical, but still not acceptable in the philosophy seminar room – would mark nominalism of the contemporary kind as involving a species of old-fashioned alienated epistemology as opposed to the "naturalized" epistemology Quine promoted.

Now, though I cannot discuss Carnap at length here, I believe that the position he held by the 1950s was not fundamentally different in this respect

[7] An arrangement of radio dish-antennas spread out over twenty miles or more, used for radio astronomy.

from that I have just associated with the name of Quine. Thus, just as Quine and Carnap and "fictionalists" are all agreed that our current science, partly owing to its being mathematically formulated, does not present a God's-eye view of the universe, so also Quine and Carnap agree that the nominalist suggestion that current science should therefore be regarded as only a "useful fiction" is to be rejected.

To put the matter another way, the "fictionalist" nominalist considers that, even when we have answered in the affirmative whether an apparatus of numbers and/or other mathematical objects does or will figure in the best physical theory, there remains a *further* question, whether numbers *really* exist, which they answer in the negative. Quine and Carnap agree in doubting that there is any intelligible question of this form.

4 QUINIANISM VS CARNAPIANISM

Thus the issue between Quine and Carnap on one side, and "fictionalist" nominalism on the other side, is over the intelligibility or appropriateness of the question, "Science and common sense aside, are there *really* numbers?" The issue between Quine on one side and Carnap on the other can also be represented as a difference over whether or not a certain question arises, or more precisely, over whether such a question as, say, "If I have as many fingers as toes, is the number of my fingers equal to the number of my toes?" arises in more than one sense. Carnap famously thought there were two senses to such a question, internal and external to the "linguistic framework" of number – or, as I will say, to the "concept" of number – where Quine took there to be only one question.

Taking the concept of number for granted – as the Carnapian is justified in doing, since the questioner has used the term "number" in framing the question – the question admits an immediate positive answer: that *if one has as many fingers as toes, then the number of one's fingers is the same as the number of one's toes*, is something that "comes with" or "is part of" the concept, and in this sense is *analytic*. This is the first, "internal" way of taking the question.

But there is another, "external" way of taking it. Perhaps in asking the question what the questioner really means to do is to raise the issue whether we should accept the concept of number. The "material mode" formulation, which has the appearance of presupposing the concept of number, may be misleading in this respect. Perhaps what is really being questioned is, put in the less misleading "formal mode," simply this: is the concept of "number" to be accepted?

This question is certainly one the Carnapian is willing to discuss, and the answer to this external question will take longer to give than did the answer to the internal question. What the Carnapian will insist is that in discussing why we accept the concept of "number," questions about the correspondence of that concept to a feature of ultimate metaphysical reality are out of order. The considerations the Carnapian advances will rather be broadly speaking "pragmatic."

Thus, for Carnap, the immediate affirmative answer is justified by appeal to linguistic considerations (by the consideration that Hume's principle or something like is analytic); by contrast, the further question why to accept that concept takes longer to answer, and the ultimate affirmative answer is justified largely by appeal to pragmatic considerations (by the consideration of the utility of mathematical formulations in scientific theory). Quine, rejecting the analytic–synthetic distinction, cannot recognize that there are two questions here.

I have said that I belong to the minority who are not so sure Quine scored a victory in his debate with Carnap; but what I want now to say is that if he *did* score a victory, it was a pyrrhic one. For in rejecting the distinction between the two ways of taking the question, Quine seems to deprive himself of any justification for giving it an immediate affirmative answer. For Quine the answer is ultimately affirmative, to be sure, but his right to give this answer seems to depend on the whole long story, involving pragmatic considerations, that has to be told to answer Carnap's *second* question. And this lays Quine open to the objection raised especially by Charles Parsons, that his account of matters cannot do justice to the felt *obviousness* of elementary mathematics. (He acknowledges the *existence* of the feeling, but has no apparatus with which to *explain* it.)

I myself consider this type of objection so serious that a Quinian ought to *want* to be able to endorse some notion of analyticity that would allow an immediate affirmative answer, "Yes, that's analytic," or "Yes, that's just part of the concept." It may be an exaggeration to say Quine *needs* a notion of analyticity if his position is to be at all plausible, since other means by which an immediate affirmative answer could be justified have not been explored; but certainly the most *obvious* means would be to accept some notion of analyticity.

(Note that "fictionalists" have no trouble returning, with their fingers crossed, an immediate affirmative answer to the question, which they will retract upon entering the philosophy seminar room. Then they will claim that all they meant was, "Yes, that's part of the usual fairy tale," or "Yes, that's how the traditional legend goes.")

5 QUINIANISM VS CARNAPIANISM, BIS

Quine, I have said, *almost* needs to accept a notion of analyticity. The question is, *can* he? To answer this question we need to look at Quine's arguments against the analytic–synthetic distinction in "Two dogmas."

And what are these two dogmas? The analytic–synthetic distinction is one. The other is the kind of logical empiricist theory of meaning according to which for each meaningful statement there must be a more or less definite range of verifying and falsifying (or, at least, of confirming and disconfirming) potential observations. The latter dogma implies the former or, at least, a theory of meaning of the kind indicated gives one way of making sense of an analytic–synthetic distinction: the *analyticity* is the limiting or degenerate case in which *every* potential observation counts in favor of the statement, and none against it. So for Quine, rejection of the second doctrine is a corollary to rejection of the first.

But Quine's writings contain also other, more independent, arguments against the second dogma, based on considerations of underdetermination of theory by evidence, the lack of crucial experiments, and the like. And whether on account of these observations or others, by 1950 many veteran logical empiricists were in the process of giving up their older theories of meaning. Difficulties with the theory had become notorious by the time of the Carnap–Quine exchange.[8]

It is therefore of some interest to consider what kind of notion of analyticity might survive rejection of the old logical empiricist theory of meaning. Perhaps the most obvious fall-back theory of meaning would be of the type philosophers of science have called "cluster concept" theories, of which Carnap's later "Ramsey sentence" view can be construed as a variant version. Let me, avoiding detailed comparisons of different views of the same general type, merely describe the kind of theory I have in mind at the highest level of generality.

The background assumption is that there must be something surveyable and graspable associated with words to guide our usage of them. On almost any conception, and certainly on Quine's, we are supposed to be able to grasp the logical form of a statement, and to grasp the basic laws of logic; for that is how we grasp the logical implications among statements.[9] On almost

[8] "Problems and changes in the empiricist criterion of meaning" (Hempel 1950) immediately followed Carnap (1950), "Empiricism, semantics, and ontology," in the same issue of the same journal.

[9] I intend "laws of logic" to be neutral here as between what would correspond in a formal system to *axioms* and what would correspond to *rules*. That there would have to be *at least one* rule is the lesson of Quine (1936).

any conception, and again on Quine's, we are supposed to be able to grasp the connection between observational terms and predicates and observable objects and properties. What remains to be considered are theoretical terms and predicates that are non-logical but also non-observational.

Here the obvious candidate for a surveyable, graspable *something* associated with an item of vocabulary would be some core theory, some basic laws. For a theoretical term generally is learnt along with a batch of related theoretical terms, and along with a batch of basic laws involving the given term and those other terms. The surveyable, graspable, body of basic laws does in at least one obvious sense guide subsequent usage of the term, and if one calls this surveyable, graspable, usage-guiding body a *meaning*, then it can be said that the basic laws that are members of that body are *part of the meaning* of the theoretical term, or *part of the concept* expressed by that term, and in that sense *analytic*.

Now already this fall-back notion of analyticity differs appreciably from more traditional notions of analyticity, and cannot do all the same work. Notably, to call something "analytic" in this sense is not at all to call it unproblematic. What is analytic must be accepted if the concept is accepted, but then perhaps the concept should *not* be accepted![10] The basic laws may, after all, be internally inconsistent, either obviously so, as with Prior's "tonk," or less obviously so, as with Frege's infamous law V.[11] Or the basic laws may have implications clashing with the results of observation. Or the term and concept may simply be otiose, creating clutter and doing no useful work.

For a Quinian, the fact that the analytic need not be unproblematic would not be an unwelcome feature. For certainly Quine wants to be able to say that there is a question whether one should accept a given concept or not, though indeed such questions are to be resolved by pragmatic considerations, and not on the basis of whether or not the concept corresponds to an ingredient of ultimate metaphysical reality. To concede that, say, Hume's principle, is analytic in the indicated sense would permit Quine to join Carnap in giving an immediate affirmative answer to the question whether the number of my fingers is the same as the number of my toes,

[10] An alternative closer to the traditional notion would be to count as part of the meaning of the term only such laws as have the form of an equation or a biconditional and the status of an abbreviatory definition. But as Quine correctly points out, the same equation or biconditional may have the status of a definition in one exposition and lack it in another.

[11] In such a case, where there is a word, a more or less definite list of basic laws, and an inconsistency in that list, should we say that there *is* a concept, but it is an inconsistent and therefore unacceptable one, or that there simply is no concept? I follow the former usage, but have found some other philosophers very strongly attached to the latter. I consider this a *purely verbal* issue in the sense of "purely verbal issue" to be discussed below.

while still doing justice to the thought that somehow its pragmatic role in scientific theorizing is relevant to the question why we regard Riemann's *On the Hypotheses Which Lie at the Foundations of Geometry* as a contribution to science, and Dickens' *Bleak House* (finished just the year before) as a contribution to art.

This rather untraditional notion of analyticity would in fact seem to be just what Quine should want, if he is to be able *both* to remain faithful to his core philosophical principle that the ultimate grounds for regarding mathematics or anything else as non-fiction rather than fiction are pragmatic, *and also* to explain the felt *obviousness* of elementary arithmetic. And yet Quine will accept neither this nor any other analytic–synthetic distinction.

And why not? Why does Quine reject the kind of theory of meaning I have just been very abstractly and very vaguely describing? Well, he does not discuss that particular kind of theory specifically, but he thinks he has reasons for rejecting *any theory of meaning at all.* He allows that if the notion of synonymy, or sameness of meaning, could be made sense of, then meanings could be admitted, being identifiable, if all else fails, with equivalence classes of expressions under the equivalence relation of synonymy. He allows that synonymy can be made sense of in terms of analyticity, and vice versa. But he famously denies that either of the two can be made sense of in a scientifically acceptable way.

But as the earliest replies to "Two dogmas" recognized, Quine in making his case is taking for granted some not-uncontroversial assumptions about what is scientifically acceptable in a linguistic or other psychological inquiry. As Grice and Strawson (1956) put it, he is assuming that only some kind of "combinatorial" or "behavioral" account could make a linguistic or psychological posit respectable. Quine's general complaint about analyticity, as applied to the particular kind of picture of analyticity vaguely sketched above, amount to just this: that there is no clear combinatorial or behavioral criterion for distinguishing which laws count as "basic" and therefore "constitutive of the meaning" of a term, and which count as "non-basic" and "additional to the meaning" of the term.

The assumption that there would have to be such a criterion before the notion could be respectable and acceptable is just an instance of the same behavioristic assumptions that notoriously led Quine to write that "different persons growing up in the same language are like different bushes trimmed and trained to take the shape of identical elephants. The anatomical details of twigs and branches will fulfill the elephantine form differently from bush to bush, but the overall outward results are alike" (Quine 1960, p. 8). The canonical objection to this assumption is given in

Chomsky's devastating review of Skinner's *Verbal Behavior* (Chomsky, 1959): children's language grows to resemble that of their parents with strikingly little input. The resemblance between the two bushes cannot be simply the result of their being trimmed the same way, because the gardeners have done very little trimming.

The conclusion is that if one is to get anywhere thinking about language, one is going to have to learn to think outside the Skinner box. The brain is not an organ for cooling the blood, and one is not going to get anywhere trying to understand the complex relations between sensory input and behavioral output simply by seeking correlations between the two, treating everything going on in-between in the brain and elsewhere as a black box. Nor can one wait until neuroscience finds the relevant physical structures inside the skull before bringing them into psychological theorizing. Without some theorizing in advance, one would not even know what to look for inside the skull, any more than, without Mendel's positing "factors" and formulating laws of heredity in terms of these, would one have known what to look for in seeking the physical basis of heredity. Theoretical posits, including meanings of a kind that bring with them a distinction between analytic and synthetic, must be allowed even if they are not directly manifested in observable behavior – provided that they play a useful role in explaining the phenomena of language and thought.

This, no doubt, most philosophers today will grant, and Quine's failure to grant it dates his classic paper more than any other feature. And yet, even granting that behaviorism of any kind, logical or substantial or methodological, is to be rejected, and that non-behaviorist programs positing meanings are to be not just tolerated but even encouraged, there is still this much to be said for Quine's skepticism about meanings: no stable consensus in favor of any one such program has as yet emerged among either linguists or philosophers. There is nothing that could be called a body of accepted scientific conclusions about meaning or analyticity that workers in other areas, such as philosophy of mathematics, can draw upon and apply to their concerns. And this being so, the question retains some interest whether a notion of analyticity can be developed *without* introducing unobservable theoretical posits, and *without* stepping outside the boundaries within which Quine confined himself.

6 QUINIANISM VS INTUITIONISM

Quine's worry about analyticity, even about the notion of analyticity sketched earlier that would seem to give him just what he should want and no more, is that there is no clear, principled way to draw the boundary between laws that are so "basic" to a term that inquirers who differ over them would be correctly described as "attaching a different meaning or concept to the term" or "speaking of different things," and laws that are not that "basic," so that inquirers who differ over them would be correctly described as "attaching the same meaning or concept to the term" or "speaking of the same things but saying different things about them." Towards indicating a way to quiet this worry, let me briefly examine some cases of apparent disagreement. I will consider cases of disagreement in logic, though examples could also be drawn from theoretical disagreements in empirical science.

At one extreme, let us consider an Australian logician who tells us that, unlike so many of his or her compatriots who merely claim that there are *some* statements of the form "P and not P" that are true, he or she maintains that *all* such statements are true. This sounds very radical. But suppose that in further conversation we find him or her frequently arguing from P or from Q to "P and Q," but never to "P or Q," and from "P or Q," but never from "P and Q" to P or to Q. This is a case where one will feel inclined to say that the radical wannabes are simply attaching a different concept to the same logical particles, meaning "and" by "or" and vice versa. And it is a case where a change of terminology would be helpful. Indeed it would actually put an end to the appearance of disagreement between Bruce or Sheila and us. That a terminological switch would thus terminate debate is the mark of a *purely verbal* dispute.

At the other extreme, consider two logicians Y and X who both accept all the simple laws of classical logic in textbooks, but who disagree over whether in a certain complicated case certain premises do or do not imply a certain conclusion. Y claims he has found a deduction showing they do. X claims she has found a counter-model showing they do not. This is a case where one will *not* feel inclined to say that the two are attaching slightly and subtly different meanings or concepts to the same words. That kind of suggestion would, in fact, sound like a bad joke. The obvious way for the two to settle their differences, given how much they have in common, would be for each to go carefully over both his or her own and the other's work, looking for a flaw in the deduction or in the model, or if necessary to bring in a third party who shares their classical orientation as referee.

For an intermediate case, consider a group of Dutch logicians who reject, if not all, then very many, instances of "P or not P," if not in the sense of denying them, then in the sense of refusing to affirm them. We may find that they persistently argue in accordance with the canons of Brouwer's or Heyting's intuitionistic logic rather than of Frege's or

Russell's or Hilbert's classical logic. In this case, sophisticated work by Gödel and others shows that for substantial parts of discourse, our way of speaking can be translated into something acceptable to them, and vice versa.[12] But there are definite limits to how far one can go, and there can be no question of translations putting a complete end to the dispute, which is therefore not *purely* verbal.

Nonetheless, one can see that, as a practical, utilitarian matter, it would be helpful for the sides to distinguish their "or"s. And in fact this is commonly done, the two "or"s being called "classical disjunction" and "intuitionistic disjunction," and similarly for negation and so forth. That there is *something* in common is indicated by "disjunction" appearing in both labels, but that there are significant differences is indicated by the different qualifying adjectives. The terminological distinction at least discourages partisans of either side from engaging in question-begging argument – "we must have P or not P, else we would have not P and not not P, which is a contradiction" – and tends to deflect controversy in the direction of meta-level discussions, which may not lead to their being very quickly settled,[13] but will at least help clarify what is at stake.

Logicians less wary of "meanings" than Quine seem spontaneously to say about this case that the meaning of "or" as used by Brouwer is different from the meaning of "or" as used by Hilbert. And as a matter of fact, Quine himself says that when the deviant logician tries to deny a classical doctrine, "he only changes the subject." The appearance of this assertion in Quine's *Philosophy of Logic*,[14] however, startled readers of his earlier works, since it seems to go clean against his whole official doctrine of repudiating meanings. What I would now like to propose is that Quinians can, without betraying the overarching Quinian principles, incorporate a notion of meaning that would make what Quine says about the deviationist not a startling lapse, but exactly right.

My proposal is that the law should be regarded as "basic," as "part of the meaning or concept attached to the term," when in case of disagreement

[12] The allusion is to the double negation interpretation, by which classical "either / or" becomes intuitionistic "not neither / nor," and to the modal interpretations, by which intuitionistic "either / or" becomes classical "it is constructively provable that either / or."

[13] Witness the immense literature spawned by Michael Dummett's neo-intuitionism, for instance.

[14] Quine (1970, p. 81). Two pages later it says that "he may not be wrong in doing so," meaning that the reasons or motives for deviations from classical logic must be examined, though Quine in fact ultimately rejects them on pragmatic grounds. Quine's position in these passages seems not far from conceding that the intuitionistically unacceptable classical laws, say, may be called "analytic" provided that term is not taken to imply "unproblematic" or "incorrigible." This contrasts with his official view in Chapter 7 of the same work.

over the law, *it would be helpful for the minority or perhaps even both sides to stop using the term or at least to attach some distinguishing modifier to it.* Such basic statements would then count as analytic.[15] This proposal makes the notion of analyticity vague, a matter of degree, and relative to interests and purposes: just as vague, just as much a matter of degree, and just as relative to interests and purposes, as "helpful." But the notion, if vague, and a matter of degree, and relative, is also pragmatic, and certainly involves no positing of unobservable psycholinguistic entities, and for these reasons seems within the bounds of what a Quinian could accept.

There is no denying that utility of the notion is limited by its vagueness, and yet I think there are some non-trivial cases that are comparatively if not absolutely clear. The intuitionist case just discussed is one of them, and I think that the case of greatest interest in the present context, Hume's principle, is another. That is to say, I think Hume's principle can be called analytic in the sense that it would be helpful if "fictionalists" would stop saying things like "I grant that if numbers exist then Hume's principle is true of them, but I don't grant that numbers exist," and instead just abandoned (inside the philosophy seminar room) the use of the term "number" (with the usual exception of use in negative existentials and in indirect discourse), and said instead, "I don't grant that use of 'number' is accepted, though I grant that if it is accepted, it should be used in accordance with Hume's principle."

What is the difference here? Well, the first formulation tends to turn discussion in the direction of the question, "Do numbers exist?" And this is a question that cannot usefully be debated between anti-nominalists and nominalists, since there is simply no agreement at all between them as to what would constitute a relevant consideration in favor of or against the statement that numbers exist. (It was in connection with this point that in my earlier paper I alluded to the interminable lawsuit *Jarndyce and Jarndyce* in Dickens.)

By contrast, much as in the intuitionist case, the second formulation tends to turn discussion in the direction of issues about what makes it appropriate or inappropriate to accept a given concept (in a philosophical as opposed to a scientific context). Though, again as in the intuitionist case, there is no guarantee that thus turning from the object level to the meta-level of discourse will result in convergence of opinion, an airing of differences at the meta-level can at least somewhat clarify *why* there seems to be

[15] As would their logical consequences, at least in contexts where, in contrast to the examples above, there is no disagreement over logic.

so little chance of achieving agreement at the object level, and debate over the criteria of acceptability of concepts does not seem as wholly futile as does debate at the object level, where it seems impossible for either side to do more than argue in a (vicious or virtuous) circle.

I will not press the point, however, but will leave it to the reader to ponder. What I want to do instead, before closing, is to consider one all-too-obvious complaint about the notion of analyticity proposed. The proposed notion of "analyticity" seems, in its relativity and vagueness, as well as in its failure to imply unproblematicity, so different from traditional ones as to make it unhelpful to use the traditional term for it, or at least unhelpful to use that term without some distinguishing modifier. Hence by my own principles I ought at least to add a qualifying adjective. Let me therefore do so, and close by commending to Quinians and non-Quinians alike a notion that may be called that of the *pragmatic analytic.*

5

Being explained away

I A LOGICIAN LOOKS AT NOMINALISM

When I first began to take an interest in the debate over nominalism in philosophy of mathematics, some twenty-odd years ago, the issue had already been under discussion for about a half-century. The terms of the debate had been set: W. V. Quine and others had given "abstract" and "nominalism" and "ontology" and "Platonism" their modern meanings. Nelson Goodman had launched the project of nominalistic reconstruction of science, or of the mathematics used in science, in which Quine for a time had joined him before turning against him. William Alston and Rudolf Carnap and Michael Dummett had raised doubts about what the *point* of Goodman's exercise could be; and though they had unfortunately been largely ignored, Quine's contention that the exercise can*not* be successfully completed had gained wide publicity as the so-called *indispensability* argument against nominalism. By contrast, two subtle discussions of Paul Benacerraf had been appropriated by nominalists and turned into the so-called *multiple reductions* and *epistemological* arguments for nominalism.

While such arguments, if sound, would suffice to establish the nominalist position even if Quine were right that mathematical entities cannot be eliminated from science, nonetheless a number of nominalists were just then setting out to prove Quine wrong. Reviving Goodman's project, but by allowing themselves means Goodman had not been willing to allow himself, they hoped to succeed where Goodman had failed: they hoped to find a way of interpreting standardly formulated scientific theories, which at least appear to imply or presuppose the existence of such things as numbers and functions and sets, in alternative theories that would not even appear to do so.

Now a lot of the work of logicians since the time of Kurt Gödel has consisted in finding interpretations of one theory in another of a superficially quite different appearance. So an experienced logician should be in a good position to give advice as to how what the nominalists were trying to do should be done. When I entered the field, I attempted just this: I undertook to tidy up the ongoing work of nominalists of the period, by indicating the *optimal* method of interpreting away numbers and sets in favor of points and regions of space-time, or interpreting away claims about the actual existence of abstract numbers into claims about the possible existence of concrete numerals. But while I was thus not impressed by claims that the nominalist project was infeasible, I was concerned over the question "Why bother?" What *I* doubted was not whether what the nominalists were trying to do could be done, but whether it was worth doing.

For logicians are used to thinking of the differences between theories that can be interpreted in each other as less important than the difference that exists when there is only an interpretation of a first theory in a second, and not the other way around. In the latter case, the second theory is overall *stronger* than the first, as logicians measure strength of theories, while in the former case the two theories are of *equal* strength. To me it was a bit surprising that so many philosophers seemed to attach great importance to differences between theories whose assumptions were of the same *overall* strength, simply because the interpretation of the one in the other involves switching what Quine called "ontological" commitments (namely, implications about what sorts of objects exist) for what he called "ideological" commitments (namely, assumptions about what sorts of predicates and what sorts of logical operators make sense). It seemed to me that this kind of difference cannot make much of a difference, because it is simply *too easy* to interpret and reinterpret and, like a "creative" accountant, move costs and benefits back and forth between the ontological and the ideological columns.

I found philosophers mostly dismissive of this attitude. They typically would suggest that a logician has the impression that it is easy to reinterpret theories to change their ontologies only because the logician has been working with theories about *abstract* entities, and that creative accounting is much more difficult when the entities with which one is concerned are *concrete*. (Indeed, this supposed difference between abstract and concrete is behind one of the standard nominalist arguments, the "multiple reductions" argument appropriated from the discussion of John von Neumann's and Ernst Zermelo's rival set-theoretic definitions of numbers in Benacerraf's "What numbers could not be.") My own impression, by contrast, was that the reason not much had been accomplished in interpreting away apparent implications or presuppositions as to the existence of concrete entities of one sort or another was that very little effort had been put into trying to do so. At the time I did not, however, attempt to show how it could be done – an omission I will begin to rectify later on in the present note. I did attempt to express my reservations and explain why I am not a nominalist, but found myself even more completely ignored than Alston and Carnap and Dummett had been.

2 REVOLUTIONARY RUMBLINGS

Meanwhile, however, Gideon Rosen was independently arriving at overlapping ideas, and a more persuasive way of presenting them; and I had the good luck that he preferred to join forces with me on a book, rather than publish his dissertation as a book on its own. Our approach in *A Subject With No Object* (Burgess and Rosen, 1997) was two-pronged.[1] We began by distinguishing, in a terminology carried over from my earlier work, two spirits in which a nominalist reinterpretation of a scientific theory might be put forward: the *hermeneutic*, according to which the nominalist version is a revelation of what the current version really meant all along; and the *revolutionary*, according to which the nominalist version is a rival to the current version intended to replace it henceforth. Rosen has since elaborated this distinction, making some subdivisions, and his elaborated form appears in a chapter in Stewart Shapiro's *Handbook of Philosophy of Mathematics and Logic* (2005). I do not want to go into great detail here, but let me review some of the key points.

Revolutionaries claim their nominalist theories are distinct from and better than current theories. But better by what standards? Revolutionaries may be subdivided into the *naturalized* (citizens of the scientific community) and the *alienated* (foreigners to the scientific community) according to whether their appeal is to scientific standards or to some supposed suprascientific philosophical standards. About the latter little need be said here, since few contemporary nominalists wish to put themselves in a position like that of

[1] Lest footnotes and bibliography become longer than the paper proper, I will refer the reader to the long list of references at the end of Burgess and Rosen (1997) for the full titles and bibliographical data of the relevant works of all the various authors alluded to in passing here, from Alston to Zermelo.

Exceptions must be made in the case of two more recent authors whom Rosen and I did not discuss in our book. One is Yablo, whom Rosen does discuss in Rosen and Burgess (2005). (My role in producing the chapter involved little more than some rewriting of Rosen's work to make it fit within the editor's word limit.) But I advise the reader to visit Yablo's website (www.mit.edu/~yablo/home.html) for the most up-to-date listing of his works, since he is actively engaged in producing new material, all of it intriguing whether one agrees with it or not.

The other exception is Azzouni (2004). My review has appeared as (Burgess 2004d).

Cardinal Bellarmine "correcting" scientific theories of planets or of Norman Malcolm "correcting" scientific theories of dreaming by appeal to the higher authority of Aristotle or Wittgenstein (as interpreted by themselves).

Generally, revolutionaries profess to be naturalized. But if they are, if they think their versions of gravitational theory or whatever are superior *scientifically* to standard versions, then one might expect them to publish their work in theoretical physics journals, or at least to attempt to do so. If "ontological economy" of mathematical apparatus really is as important to scientists as it is to certain philosophers – something I myself very much doubt, since it is *very* difficult to find any clear historical instance of such a preference – such contributions ought to be welcome. Yet the experiment of submitting a write-up of a nominalist project to a theoretical physics journal has never been tried, so far as I know, and candid revolutionaries of a professedly naturalist stamp would probably concede that if undertaken, the test would very likely be failed: the papers would *not* make it through peer review. But if this is admitted, how *can* a revolutionary profess to be a naturalist, adhering to scientific standards in judging theories?

One common line is to claim that though nominalist physics is perhaps not superior to mathematical physics by the standards of physicists, what really need to be compared are not just the two versions of physics, but rather two packages of combined physics and epistemology. Somehow nominalist revision, which may make the job of the physicist more difficult, is supposed to make the job of the naturalized, scientific epistemologist easier. In what way? It is at this point that nominalists of the school I have been alluding to bring forward their appropriation of Benacerraf's discussion of knowledge in his famous paper "Mathematical truth." The so-called Benacerraf problem is the puzzle, "How could we come to know anything asserting or implying or presupposing that there are numbers or functions or sets, given that it does not make sense to ascribe spatiotemporal location or causal powers to such mathematical entities?" Nominalism provides a very easy answer to this "How can we?" question – namely, the answer "We can't!" – which otherwise would be a difficult one, it is said.

This line of thought involves a serious confusion, which can be brought out by considering what properties a belief must have in order to rank as knowledge. These are three: justification, truth, and whatever it takes to bridge the gap between justified true belief and knowledge that was discovered by Edmund Gettier. But the epistemological argument for nominalism is not about Gettierology. Nor is it really about truth. (The nominalist argues that standard mathematical existence theorems cannot be *known* to be true as a way of avoiding directly confronting the question of whether they *are* true.) So the issue is one of justification. Once this is appreciated, it can be seen that the whole idea of trading costs to physics against benefits to epistemology is a muddle. For providing an explanation of the historical fact that current mathematical and scientific theories have come to be believed will be an important task for scientific, naturalized epistemology *regardless* of whether or not one takes belief in those theories to be justified. This task is in no way made easier by the assumption that the belief in question is *un*justified.

The obvious *anti*-nominalist solution to the Benacerraf puzzle is to suggest that if you cannot think how we could justifiably come to believe anything implying, say, "There are functions," then just look at how mathematicians come to believe, say, Gödel's result, "There are solutions to the field equations of general relativity with closed time-like paths." *That's* how one can justifiably come to believe something implying "There are functions." The revolutionary nominalist who rejects this answer must think the actual historical process leading to belief in this theorem of Gödel's is somehow not a *justifiable* process of belief-formation. But it is virtually a tautology that the belief arrived at is justifiable *by mathematico-scientific standards*. And hence the revolutionary nominalist's position, according to which it is *un*justifiable, must involve covert appeal to suprascientific philosophical standards of justification – must be alienated – after all.

To vary the example, let us consider the claim of nominalists who maintain that what they are appealing to is just "what science teaches us about how we humans obtain knowledge," and see how this applies to, say, the belief that more than a half-dozen books advocating nominalism have been published in the past three decades or so. By "books" here I clearly do not mean concrete book *tokens*, since there are not just "more than a half-dozen" but hundreds or thousands of such tokens, scattered through various institutional and personal libraries. So the belief in question is one about abstract book *types*, and hence according to the nominalist must be something "science teaches us" we cannot know. But is this a teaching of *science*, or of some Procrustean epistemological theory? If you asked me for evidence to justify the belief that more than a half-dozen books advocating nominalism have been published in the past three decades or so, I could point to various book tokens on the shelves of my office, with titles like *Ontology and the Vicious Circle Principle* and *Science without Numbers* and *Mathematics without Numbers*, and names like "Hartry H. Field" and "Charles S. Chihara" and "Geoffrey Hellman" on the title page, along with dates like 1973 and 1980 and 1989. Can anyone seriously maintain that *science* teaches us that this is insufficient evidence?

3 HERMENEUTICAL HI-JINKS

Turning to hermeneutic nominalism, the most obvious objection to its claim that what appear to be statements about numbers and sets are really statements about something quite different, is the simple lack of *evidence* for it. But there is another problem, which can be illustrated by the case of the proposal to paraphrase away apparent talk of the existence of abstract numbers as really being talk of the possible existence of concrete numeral-tokens. The hermeneutic nominalist who resorts to this kind of paraphrase – and similar remarks would apply to those who favor other kinds – will want to say, as the revolutionary nominalist does not, that "There are prime numbers greater than 10^{10}" is true, justifiably believed, and so on, because "deep down all it really means" is something like "There could have existed prime numeral-tokens greater than 10^{10}." The trouble is that parity of reasoning suggests then that "There are numbers" must equally be true, justifiably believed, and so on, because "deep down all it really means" is something like "There could have existed numeral tokens." But whether "There are numbers" is true, justifiably believed, and so on, *was the whole original issue.* Certainly this was the question Goodman and Quine were asking when they first agitated the issue of nominalism (and not, for instance, some question about hypothetical "deep structures," in which neither Goodman nor Quine ever believed). To concede that "There are numbers" is true, justifiably believed, and so on, is to concede all the anti-nominalist maintains. (Perhaps anyone who really deserved to be called a "Platonist" in any historically serious sense would want to claim more; but I doubt that there are any living Platonists in any such sense of "Platonist.")

Such, in brief, were the kinds of arguments Rosen and I put forward in *A Subject With No Object.* Since the appearance of that book it has become apparent, however, that hermeneuticists, like revolutionaries, are divisible into two subcategories, which Rosen has called *content* hermeneuticism and *attitude* hermeneuticism. The former is the kind of view I have been discussing so far, about what mathematically formulated statements "deep down really mean." The latter is not a view about meaning in this sense, but about the *attitude* of mathematicians and scientists and the lay public towards scientific and mathematical and commonsense theories apparently involving abstract entities. Attitude hermeneuticism is the view that, contrary to the common assumption of the anti-nominalist, revolutionary nominalist, and content-hermeneutic nominalist, such theories *are not really believed.* As developed by Steve Yablo and others, the attitude-hermeneutic view has been the dominant version of nominalism over the better part of the past decade, though the attitude-hermeneuticist line has zigged and zagged a bit over the course of that period.

First we were told mathematics is like fiction. Well, it is not, and this in two crucial respects. For one thing, our *attitudes* towards mathematics and towards fiction are totally different: we *rely* on mathematics in important practical applications, as we do not rely on novels, short stories, and the like. (If we need the services of a good detective, we do not go to Baker Street.) For another thing, when there is a dispute about whether some particular text is fiction or non-fiction – the Don Juan books of the elusive Carlos Castañeda may serve as an example – we at least have a pretty clear idea what it would be *like* for it to be true or not true. For instance, we have a pretty clear idea of what it would be like for a man to smoke a concoction of dried mushrooms, turn into a bird, and fly off a cliff. By contrast, once we join the nominalist in abandoning ordinary mathematical standards for judging when mathematical existence claims are true, or have been adequately proved, we are left with no other agreed standards.

Then we began to be told that mathematics is like metaphor or some related figure of speech. Well, again it is not. For one thing, as Rosen argues in our chapter in Shapiro (2005), metaphorical usages can almost always be instantly recognized by the speaker as having been meant non-literally, as soon as anyone raises the issue, whether or not one is able to say in literal terms what *was* meant; and again that is far from being true in the mathematical case. The "figuralist" or "figurative" interpretation seems to be attributing to mathematicians and scientists and lay people too philosophically sophisticated an attitude.

I do not know what we will be told next. Fictionalism and figurativism do not exhaust the options for the attitude hermeneuticist – in ongoing work Yablo has some very interesting things to say about "non-catastrophic presupposition failure" – though I think it is clear by now that attitude hermeneuticism is not something arrived at by first studying some linguistic phenomenon, then noticing that the conclusions one draws about it have nominalistic implications. Rather, a commitment to nominalism seems to be there first, and to be what is driving the search for some linguistic phenomenon or other whose analysis could somehow or other be applied to support the nominalist position.

Setting Yablo's developing views aside, let me turn to the latest book-length work on the issue of nominalism, Jodi Azzouni's *Deflating Existential Consequence* (2004). This work makes the mind-boggling claim that one can sincerely assert "There are such things as numbers" and even "'There are such things as numbers' is literally true" and *still* not be "ontologically committed" to numbers. Azzouni's position may perhaps be classed as an extreme version of attitude hermeneuticism, but owing to its extremism it is perhaps best put in a class by itself. However it ends up being classified, there is clearly something radically wrong with it.

In the first place, to repeat an earlier observation, whether it is true that there are numbers was *the whole issue*, and in conceding that it *is* true, the would-be nominalist of this new style is conceding everything the anti-nominalist maintains. In the second place, the claim about "ontological commitment" that the new-style would-be nominalist is making is self-contradictory. For "ontological commitment" was a phrase without use and therefore without meaning until Quine gave it a meaning by stipulative definition; and that stipulative definition makes sincere assertion that there are numbers, or that "There are numbers" is literally true, a more than sufficient condition for "ontological commitment" to numbers.

In an effort to make some kind of sense of Azzouni's nonsensical claim I was led to speculate that what he has done has been to take Quine's phrase "ontological commitment" and substitute for Quine's understanding of "ontological," on which the word is merely a fancy synonym for "existential," some other understanding of "ontological," presumably adopted from some pre-Quinean tradition. So let me, without making any strong exegetical claim about Azzouni, examine the contrast between pre-Quinean and post-Quinean senses of "ontology" and its derivatives. For this purpose we must turn back to a time before the beginning of the debate on modern nominalism, and I think we do best to turn quite a way back, right back to the beginning of the modern era.

4 READING GOD'S MIND OR IMPOSING A SCHEME ON THE WORLD?

My account of the history will be condensed to the point of being a cartoon, but nonetheless I hope it may help the woods stand out from the trees. I begin with a much-quoted passage in William James, describing the attitude of the heroes of the Scientific Revolution, who hoped for a science that would be nothing less than a reproduction in our minds of the blueprint for the universe used by the Great Architect:

> When the first mathematical, logical, and natural uniformities, the first *laws*, were discovered, men were so carried away by the clearness, beauty and simplification that resulted, that they believed themselves to have deciphered authentically the eternal thoughts of the Almighty. His mind also thundered and reverberated in syllogisms. He also thought in conic sections, squares and roots and ratios, and

geometrized like Euclid. He made Kepler's laws for the planets to follow: he made velocity increase proportionally to the time in falling bodies; he made the law of the sines for light to obey when refracted; he established the classes, orders, families and the genera of plants and animals, and fixed the distances between them. He thought the archetypes of all things, and devised their variations; and when we rediscover any one of these his wondrous institutions, we seize his mind in its very literal intention. (James 2000, p. 29)

To show James is not just making this up, I could reproduced much-quoted passages from Galileo's *Assayer* and Kepler's *Astronomia Nova*, but let me forbear.

The goal for those who accepted this picture was to produce a description of reality "just as it is in itself," or equivalently a description of the universe as God sees it, and not as we see it. (Take the invocation of the Deity literally or metaphorically as you choose.) Such a description would necessarily be very different from the description of our environment which we use in everyday life. (According to the seventeenth-century worthies I have been alluding to, a chief difference would be that the colors and sounds and odors we see and hear and smell would be gone, and only size and shape and position, along with speed and direction of motion, would be left.)

But as David Hume already saw, if one makes one's standard for "knowledge" the possession of a representation of reality that describes it "just as it is in itself," then the consequence will "an universal skepticism" and the conclusion that "knowledge" is impossible. Hence Immanuel Kant's Copernican revolution. For Kant, the aim is still to separate out, in our ordinary and scientific accounts of the world, what is contributed by the world and what by us; but instead of attempting to do this by producing an account with *nothing* contributed by us, Kant proposed to proceed the other way around, by producing an account with nothing contributed *by the world*, an account of the pure forms of sensibility and categories of the understanding supplied by us, into which the world pours empirical content.

In the century and a half between Kant and Carnap, which I will leap over in a single bound, there was really surprisingly little change in the nature of the project. With Carnap there is more talk of "linguistic frameworks" and less of "pure forms of sensibility" or "categories of the understanding," and there is a shift from claims about what we *inevitably must* impose to claims about what we *conventionally do* impose on the world. But even if for Carnap there is no one conceptual scheme we must impose on the world, yet still we must impose some conceptual scheme or other, and there can be no question of getting behind any and every conceptual scheme to the world "just as it is in itself." Alongside this agreement of substance between Kant and Carnap, there is a disagreement over terminology, and in particular over the role of the term "metaphysics." Originally this term applied to the attempt to get behind our conceptual schemes to a God's-eye view of reality, something Kant and Carnap agree is impossible. Kant proposes to use it instead for his own project of articulating just what the scheme our intuition and understanding impose on the world amounts to. Carnap, by contrast, proposes simply to retire the term. Thus for Kant "the future of metaphysics is critique," while for Carnap metaphysics has no future.

Against Carnap, Quine claimed that while the fabric of our theory of the world is "white with convention" and "black with fact," there are no purely black threads and no purely white threads in it. The point about *black* had, in effect, already been conceded, or rather insisted upon, by Carnap when he argued, *contra* Moritz Schlick, that the evidence in science consists of corrigible reports of observations about the furniture and implements of the laboratory, and not incorrigible reports about sense-data. The point about *white*, about the existence of a purely conventional element, was the issue between Carnap and Quine.

Quine's contention was just this. Suppose, as Carnap maintains, we generally favor one linguistic framework or conceptual scheme over another on grounds of convenience: in attempting to describe the world, we find it better suits our purposes to do so in using *this* framework or scheme rather than *that*. Well, what sort of fact is this fact that one scheme is more convenient than another for us to use in attempting to deal with the world? It would seem to be a fact not just about us, but also about the world: *we* are such and *it* is such that *we* can more successfully deal with *it* in this way rather than that. So the scheme is not, after all, something contributed purely by *us*, since part of the reason we choose it is that *the world* lends itself to description in terms of these conceptual resources rather than others.

Rightly viewed, the difference between Quine and Carnap here is one of detail: much more unites than divides them. In particular, Quine has no more use than Carnap for the kind of pre-Kantian project of attempting to describe reality "just as it is in itself." And yet there is a terminological difference between the two over the term "ontology," traditionally a near-synonym for "metaphysics." Quine agreed with Carnap that ontology in this traditional, pre-Kantian sense is meaningless. Quine, however, differed from Carnap in what he called the "ethics of terminology," insisting that if a word was meaningless, he had the right to give it a meaning by stipulative definition, and choosing to exercise this alleged right in the case of the word "ontology." The new enterprise of "ontology" in the post-Quinean sense is simply a glorified taxonomy, an attempt to catalogue what sorts of objects there are in reality, not "just as it is in itself" but as apprehended by us through our everyday and technical language, our commonsense and scientific theories.

This untraditional use of "ontology" is of a piece with the historically dubious use of "nominalism" and the historically absurd use of "Platonism." (In any traditional sense, it is people James is talking about, people like Galileo and Kepler, who are the Platonists, while an anti-metaphysical pragmatist like Quine is no more a Platonist than was James.) Why Quine chose to apply an old label to a new project is to me something of a mystery. It is clear that having a synonym, "ontological" or "ontic," for "existential" must have been useful during the heyday of Jean-Paul Sartre. Readers would have winced if the section of *Word and Object* entitled "ontic decision" had instead been entitled "existential choice." I fear, however, that Quine may have chosen to use "ontology" mainly to needle Carnap, who seems to have more than just disliked the word (perhaps because it was a favorite of Heidegger, who incidentally also used it in a radically untraditional sense). The danger posed by Quine's transferring "ontology" from the old project to the new – rather than coining contrasting labels – is that some may be led to confuse the two homonymous enterprises.

And just this is what I suspect may have happened in the case of those recent nominalists who say in one breath, "I sincerely believe that it is literally true that there are such things as numbers," and in the next, "I am in no way ontologically committed to numbers." This otherwise nonsensical double-talk becomes less nonsensical if one takes "ontology" in the second assertion to be meant in a pre-Kantian rather than a post-Quinean sense. Indeed, while I myself sincerely believe that it is literally true that there are such things as numbers, I do not believe that the aim of traditional, pre-Kantian ontology (namely, the aim of getting behind all conceptual schemes to reality "just as it is in itself" and cataloguing what sorts of objects it contains) is a feasible one. Of course, this being my attitude, I wish to make "ontological" claims, in a traditional, pre-Kantian sense, *neither* for abstract objects *nor* for concrete ones. It is here that I differ from what seems to be the attitude of the double-talking nominalists, who go on to say in their third breath, "But I *am* ontologically committed to this table and these chairs, and to the moon and the stars." What I see wrong in this kind of nominalism is not its "anti-realism" about the abstract, but what appears to be its "realism" (in a traditional, pre-Kantian sense) about the concrete.

5 ABSTRACT SKEPTICISM VS CONCRETE CREDULITY

What I am inclined to conclude from the tendency observable over these last decades for nominalism to morph from one form to another is that nominalism can never be defeated by arguments solely about the abstract, since what feeds it is an underlying naïveté about the concrete and our knowledge thereof. It is for this reason that I welcome recent epistemological arguments for what I will call – from the Greek word for "simple" – the *haplist* position. As the nominalist holds that everything there is is concrete, and hence that there are no numbers, no books (in the sense of types rather than tokens), and so on, so the haplist holds that everything there is is simple, not extended and composite, and hence that there are no chairs and tables, and no moon and stars – and no people, and in particular no haplist philosophers! Though the haplist conclusion is absurd, attention to what haplists have to say may at least help show that the explanation of our knowledge of the concrete is not so straightforward as nominalists seem to suppose. This is especially so since the form of the epistemological argument for haplism is so similar to that of the epistemological argument for nominalism.

The nominalist's skeptical argument goes something like this: I look at my hand and see that (counting the thumb as a finger) there is a finger and another and another and another and another and no more, and conclude that the number of fingers on my hand is five. But if we look at what fundamental physics tells us is really going on here, what we find is just this: light coming from an external source is reflected off my fingers over there to my eye over here, beginning a process in my body that ends with my forming the belief that the number of fingers on my hand is five. But assume what you will about whether, in addition to the concrete fingers, such an abstract entity as the number five exists or not, no such alleged thing plays any role in this explanation. If I end up speaking as if there were such a thing, there actually being such a thing plays no role in explaining why I do: the explanation why I do must be sought quite elsewhere, in the convenience of positing such "useful fictions" as numbers and sets for purposes of getting on in the world.

The haplist's skeptical argument goes rather like this: I look over there and see something brown and chair-shaped, and conclude that there is a chair over there. But if we look at what fundamental physics, as in Richard Feynman's *Q.E.D.*, tells us is really going on here, what we find is just this: photons coming from an external source are absorbed by the electrons among the myriad fundamental particles swarming in chair formation over there, some of which electrons quickly emit other photons directed over here, initiating a process – and so on. But assume what you will about whether, in addition to the simple fundamental particles, such an extended, composite entity as the chair exists or not, no such alleged thing plays any role in this explanation, in which the electrons and quarks do all the work. If I end up speaking as if there were such a thing as the chair, there actually being such a thing plays no role in explaining why I do: the explanation why I do must be sought quite elsewhere, in the infeasibility of my keeping track of the complex motions of the myriad tiny fundamental particles, and consequent convenience of positing such "useful fictions" as chairs and tables for purposes of getting on in the world.

Pointing to the parallelism between the two forms of skepticism, I submit that if the haplist's is nothing more than a clever sophism (as I imagine most nominalists would agree it is), then the nominalist's is no better. Still, I would like to make a stronger case against the claim that while ultimate metaphysical reality "as it is in itself" does *not* contain numbers or books, by contrast it *does* contain tables and chairs, or the moon and the stars. This brings me at long last to the topic my title was intended to herald: the reasons for doubting that ultimate metaphysical reality "as it is in itself" contains objects *of any sort*. These reasons were adumbrated in a section of *A Subject With No Object* that has been little read, but I think it is time to refresh and elaborate upon the suggestion made there, in the hopes of moving the never-ending but ever-changing debate over nominalism in a new direction.

6 TALKING OF OBJECTS – OR NOT

We speak of the world as containing objects with properties inasmuch and insofar as we speak a language with nouns and verbs, and sentences with subjects and predicates. The position to which I subscribe and that I wish to defend here is that there is no reason to suppose, just because we speak to each other in such a language, that God speaks to himself in such a language, or that the object-property structure is a feature not merely of reality as apprehended by us, but of reality as apprehended by God, or equivalently, "as it is in itself." There is nothing *wrong* with our speaking as we do, and certainly I do not myself propose to speak otherwise, but there is nothing *uniquely* right about it either, and if other intelligent creatures do not do so, they are not necessarily making a mistake.

Now as a matter of fact, though I have said that we speak to each other in a language with certain grammatical features, it is not beyond controversy that all human languages do in fact share these features. Some linguists have claimed otherwise, as in the following passage from Whorf:

[I]n Nootka, a language of Vancouver Island, all words seem to us to be verbs, but really there are no classes [of nouns and verbs]; we have, as it were, a monistic view of nature that gives us only *one* class for all kinds of events. "A house occurs" or "it houses" is the way of saying "house," exactly like "a flame occurs" or "it burns." These terms seem to us like verbs because they are inflected for durational and temporal nuances, so that the suffixes of the word for house event make it mean long-lasting house, temporary house, future house, house that used to be, what started out to be a house, and so on. (Whorf 1956, pp. 215–16)

And some literary writers have imagined a whole world of speakers of such languages, as in the following passage from Borges:

Hume noted once for all time that Berkeley's arguments did not admit the slightest refutation nor did they cause the slightest conviction. This dictum is entirely correct in its application to earth, but entirely false of Tlön. The nations of this planet are congenitally idealist. Their language and the derivations of their language – religion, letters, metaphysics – all presuppose idealism. The world for them is not a concourse of objects in space; it is a heterogeneous series of independent acts. It is successive, not spatial. There are no nouns in Tlön's conjectural *Ursprache*, from which the "present" languages and dialects are derived: there are impersonal verbs, modified by monosyllabic suffixes (or prefixes) with an adverbial value. For example: there is no word corresponding to the word "moon," but there is a verb which in English would be "to moon" or "to moonate." "The moon rose above the river" is *hlör u fang axaxaxas mlö*, or literally: "upward behind the on-streaming it moon[at]ed." (Borges 1962, p. 23)

(In what follows let me use "moonate" – or perhaps better, "lunate" – since the other verb suggested already exists in English in a vulgar sense.)

Whorf is speaking about an actual language, and if he is right, then a noun-free language is not only possible but actual. Unfortunately, however, though Whorf is speaking of real people, it has been disputed whether what he is saying about them is really true. Borges, of course, is only describing a fictional planet. That does not matter for us philosophers, since all we are interested in is the *possibility* of speaking a language without nouns. But Borges does not really show this, since his description of the language of Tlön does not go into enough detail to convince one that the bulk of the things we might like to say could be replaced by saying things using only verbs and adverbial modifiers.

What I wish to review here is a different approach to showing how a language like English could be translated into a language with only those grammatical categories. Of course, if one's *only* understanding of this new language were by way of explanations of how to translate it into English or English into it, no conclusions about "ontology" would follow; so an effort of the imagination is still required to convince oneself that children could grow up being spoken to and speaking such a language and no other. I trust this will not be too difficult, but ultimately the reader must judge.

To begin with, we need to imagine English translated or, as Quine called it, "regimented" into what logicians call a first-order language (with predicates only and no singular terms). The possibility of such regimentation is what lies behind Quine's slogan "to be is to be the value of a variable." Here I must assume familiarity, from Quine's writings or elsewhere, with how such regimentation might be attempted. To give at least one example, consider the following truth:

(1) Whatever lives, changes.

Now (1) can be regimented as follow:

(2) $\forall x(x \text{ lives} \rightarrow x \text{ changes})$

Also (2) admits several equivalents, including one involving only negation, conjunction, and existential quantification:

(3) $\sim\exists x(x \text{ lives} \wedge \sim(x \text{ changes}))$

In his paper "Variables explained away," Quine shows how we can eliminate variables like the x in (3). We first enrich our language with new operators, the so-called *predicate functors*, operators that attach to predicates to form new predicates, defined thus:

$(\nu F)x_1 \ldots x_m$	$\sim Fx_1 \ldots x_m$
$(\kappa FG)x_1 \ldots x_m y_1 \ldots y_n$	$Fx_1 \ldots x_m \wedge Gy_1 \ldots y_n$
$(\sigma F)x_1 \ldots x_{m-1}$	$\exists x_m(Fx_1 \ldots x_{m-1}x_m)$
$(\rho F)x_1 \ldots x_{m-1}$	$Fx_1 \ldots x_{m-1}x_{m-1}$
$(\phi F)x_1 \ldots x_{m-1}x_m$	$Fx_1 \ldots x_m x_{m-1}$
$(\psi F)x_1 \ldots x_{m-1}x_m$	$Fx_m \ldots x_{m-1}x_1$

Going from English to Quinese, each expression in the right-hand column may be abbreviated by the corresponding expression in the left-hand column. Writing for short "F" and "G" for "lives" and "changes," (3) becomes:

(4) $\sim\exists x\,(Fx \wedge \sim Gx)$

It can then be reduced to an equivalent without variables in the following steps

(5) $\sim\exists x\,(Fx \wedge (\nu G)x)$
(6) $\sim\exists x\,(\kappa F(\nu G))xx$
(7) $\sim\exists x\,(\rho\,(\kappa\, F(\nu\, G)))x$

(8) $\sim(\sigma\ (\rho\ (\kappa\ F(\nu\ G))))$
(9) $\nu\ (\sigma\ (\rho\ (\kappa\ F(\nu\ G))))$

Going back and restoring "lives" and "changes" for "F" and "G" in (9) we have

(10) ν (σ (ρ (κ lives (ν changes))))

My personal contributions in this area have been two. First, I thought of combining Quine's slogan "to be . . ." with his paper title ". . . explained away" to produce the title of the present paper. Second, I suggested a way of *pronouncing* the predicate functors:

(ν talks)*x*	*x doesn't* talk
(κ walks runs)*xy*	*x* and *y respectively* walk *and* run
(σ stares at)*x*	*x (just)* stares
(ρ destroys)*x*	*x self*-destructs
(φ eats)*xy* or (ψ eats)*xy*	*x suffers* or *undergoes* eating by *y*

Applying this suggestion to (10), we can go back from symbols to words in the following steps:

(11) ν (σ (ρ (κ lives (doesn't change)))
(12) ν (σ (ρ (respectively live and don't change)))
(13) ν (σ (self-respectively lives and doesn't change))
(14) ν (just self-respectively lives and doesn't change)
(15) doesn't just self-respectively live and not change

Here we have a noun-free verb phrase. We may, if we wish, supply a subject, "The Absolute," or we may indicate that the verb phrase is a complete sentence in itself by using the obsolete third-person singular verbal ending -th, as when one eliminates the pleonastic subject pronoun in "it rains" by writing, as Quine somewhere suggests, "raineth." We thus have the choice between two options:

(16a) *Monist*: The Absolute doesn't just self-respectively live and not change.
(16b) *Nihilist*: Doth not just self-respectively live and not change.

Two subsidiary points should be emphasized. First, one needs some way not merely of making assertions, but also of carrying out *arguments* in the new kind of language – some way other than translating back into first-order terms and applying text-book rules. In fact, John Bacon and others have supplied proof procedures, which, however, cannot be gone into here. Second, the point noted by Johann van Benthem should be emphasized, that if one starts with a *many-sorted* first-order language, one can apply the tricks I have been describing to some of the sorts and not the rest of them, retaining whatever sorts of objects one likes, and eliminating whatever sorts of objects one does not.

7 THE DARK SIDE

Thus whether one speaks overtly of abstract objects or concrete objects, of simple objects or compound objects, or indeed of any objects at all, is optional. My claim is that if children who grew up speaking and arguing in Monist or Nihilist or some Benthemite hybrid between one or the other of these and English, it would be gratuitous to assume that covertly they are "committed" to a full range of sorts of objects, just as if they spoke a language like ours, with a full range of sorts of nouns. And any assumption that the divine *Logos* has a grammar more like ours and less like theirs would be equally unfounded, I submit. It is in this sense that I claim any assumption as to whether ultimate metaphysical reality "as it is in itself" contains abstract objects or concrete objects, or simple objects or compound objects, or again any objects at all, would be gratuitous and unfounded.

This kind of anti-metaphysical claim, if not quite the kind of reason for it that I have offered, has been characteristic of pragmatism from James onward. I have mentioned one recent pragmatist, Quine, whom I believe to hold essentially this sort of view, despite his very regrettable coquetting with the modes of expression peculiar to early modern metaphysicians. I need now to say something about another recent pragmatist, Hilary Putnam.

Putnam is often cited alongside Quine as the second author of the indispensability argument against nominalism, but as explained in *A Subject With No Object*, there is an important difference between Putnam and Quine here. This is because when Putnam put forward his indispensability argument he had already committed himself to a doctrine of "equivalent descriptions," according to which there is nothing to choose between a conventional formulation of mathematics in set-theoretic terms and an alternative formulation in modal-logical terms. Thus what he was really claiming to be indispensable for science was something of the *overall* strength of classical mathematics, as opposed to constructive mathematics of one kind or another. He was not making any claim about the indispensability of *ontological* commitments (to sets) specifically, since he thought these could always be traded for *ideological* commitments (to modality). From a logician's point of view, Putnam's claim is a good deal more interesting than Quine's, but this is not the place to go into that aspect of Putnam's views.

An aspect that does require discussion is Putnam's very regrettable coquetting, like James before him, with the modes of expression peculiar to traditional idealist metaphysicians, rather as another recent pragmatist, Richard Rorty, coquettes with the modes of expression peculiar to contemporary *post-modernes.* I am alluding to the tendency to say that the moon and the stars are "mind-dependent," or worse, "socially constructed." There is something right in what Putnam maintains, and even in what Rorty maintains, and my hope is that my sketch of an alternative kind of language can help us separate this correct element from pernicious nonsense about mind dependence and social construction – really amounting to little more than what Quine called a "use-mention confusion," with or without the added twist of a confusion of academic radical skepticism with genuine political radicalism – emanating from the idealist or po-mo Dark Side.

The view from the Bright Side is that if we do choose the conventional option, and follow the conventional rules for making and evaluating claims about objects, we must conclude that the moon and the stars long antedate human mentality and society, and therefore cannot be dependent on the former and cannot have been constructed by the latter. On the other hand, if we choose an alternative option, then we will not be speaking about objects at all, and among the objects of which we will *not* be speaking or saying anything will be the moon and the stars, and among the things we will *not* be saying about them is that they are mind-dependent or socially constructed. One either plays the language game by the rules, or does not play it all, and in neither case is saying "The moon is mind-dependent" or "The stars are socially constructed" a legitimate move. We may choose between "moon" and "stars" on the one hand and "lunate" and "stellate" on the other, but if we take the first option, we must say that the moon and the stars were there long before there were astronomers or human beings or primates or mammals or animals or life, while if we take the latter option, we must say that the Absolute was lunating and stellating long before it began to astronomize or humanize or primatize or mammalize or animalize or vitalize.

All this is merely by way of avoiding certain objections to pragmatism resulting from injudicious diction on the part of some of its most distinguished advocates. On issues of substance rather than style, I stand with the pragmatist tradition: I agree with James that "the trail of the human serpent is over all," and think that *this* – and not some lesson about the impossibility of mathematical knowledge – really is something "science teaches us about how we humans obtain knowledge." From the pragmatist thesis that it is impossible for human beings to obtain a God's-eye view of the world, I infer the anti-nominalist corollary that it is pointless to complain that, for all we can know, mathematical objects may not be part of the world as God sees it. To be sure, I am well aware that the considerations I have presented above are very far from constituting a knock-down argument for that anti-nominalist conclusion. But what the course of the debate over nominalism seems to me to reveal is that arguments are not what are needed at this point: nominalists are in the grip of a picture, and that until that grip is shaken, no argument, however cogent, can hope to accomplish more than to cause nominalism to morph once again into some new form: knock down one form of nominalism, and another will pop up in its place. What is really needed is a *Gestalt* switch, and this the above sketch of another way of speaking may perhaps help to induce.

6

E pluribus unum: *plural logic and set theory*

I THE ORIGIN OF SET THEORY

If one is interested in how best to formulate and motivate axioms for set theory, it is worthwhile to take another look at the early history of the subject, right back to the work of its founder, Cantor. Cantor's definition of a set was "any collection into a whole" of "determinate, well-distinguished" sensible or intelligible objects. According to a well-known quip of van Heijenoort, this definition has had as much to do with the subsequent development of set theory as Euclid's definition of point – "that which hath no part" – had to do with the subsequent development of geometry. But in fact the notion of a many made into a one, which is what Cantor's definition makes a set to be, will repay some study.

In order to give concrete meaning to Cantor's abstract definition, our study should begin with a look at the context in which Cantor first felt it desirable or necessary to introduce the notion of set. As is well known, Cantor's general theory of arbitrary sets of arbitrary elements emerged from a previous theory of sets of points on the line or real numbers. This itself emerged from work on Fourier series. The technical details of Cantor's theorems on this topic are irrelevant for present purposes, but the general *form* of his results should be noted.

Cantor's first result was a certain (uniqueness) theorem for a given series, which depended on the assumption that at every point or number x without exception, a certain (convergence) condition holds. He then was able to relax this assumption, and show that the theorem still holds if there is only one exceptional point, or only two, or only three, and so on. A more substantial generalization was that the theorem still holds even if there are infinitely many exceptional points, provided they are *isolated* from each other, in the sense that for every exceptional point there are two points such that it is the only exceptional point lying between them.

In connection with this last result and further generalizations, Cantor introduced a new notion. If X is a set of points, let ∂X be the set of points in X that are not isolated from other points in X. Then writing E for the set of exceptional points, and writing $\emptyset$ for the empty set, the last result mentioned may be restated as saying that the theorem still holds if the following condition is met:

(1) $\partial E = \emptyset$.

Cantor was then able to prove that the theorem still holds if ∂E has only one element, or only two, or three, and so on, or even if there are infinitely many points in ∂E, provided they are isolated from each other.

Writing $\partial^2 X$ for $\partial(\partial X)$, this last result says the theorem still holds if the following condition is met:

(2) $\partial^2 E = \emptyset$.

Similarly, a sufficient condition would be $\partial^3 E = \emptyset$, or $\partial^4 E = \emptyset$, and so on. Writing $\partial^\omega E$ for the intersection of all ∂^n for $n = 1, 2, 3$, and so on, the following is also a sufficient condition:

(3) $\partial^\omega E = \emptyset$.

Cantor went on to iterate the operation ∂ beyond ∂^ω, introducing in the process the transfinite ordinal numbers $\omega + 1$, $\omega + 2$, and so on, but we need not follow him further.

At what stage in this process of generalization does it become indispensable to think of the infinitely many exceptional points as together forming a single object, a set E to which an operation ∂ can be applied and reapplied? Close examination of the form of the theorems shows that for any fixed finite n the condition $\partial^n E = \emptyset$ can be expressed as a condition mentioning only the exceptional points, and not the set E or the operation ∂, while beyond this stage, restatement in terms of just the points and not the point-set is impossible.

Consider, for instance, the hypothesis that (1) fails but (2) holds, or in other words the following:

(4) $\partial E \neq \emptyset$ but $\partial^2 E = \emptyset$.

Without mentioning E or ∂, this can be expressed, as follows:

(5) Not every exceptional point is isolated from all other exceptional points, but every exceptional point that is not isolated from all other exceptional points is isolated from all other exceptional points that are not isolated from all other exceptional points.

If we introduce a predicate E*x* for "*x* is exceptional," and use the usual symbol $<$ for order, and if we spell out explicitly the definition of isolation, (5) can even be restated in the formalism of first-order logic, with variables ranging only over points, as follows:

(6) $\sim\exists u(\mathrm{E}u \rightarrow \mathrm{I}(u))\ \&\ \forall u(\mathrm{E}u\ \&\ \sim\mathrm{I}(u) \rightarrow \exists v_1\exists v_2\forall v(v_1 < v\ \&\ v < v_2\ \&\ \mathrm{E}v\ \&\ \sim\mathrm{I}(v) \leftrightarrow v = u))$

wherein I(u) is an abbreviation:

$\mathrm{I}(u) \leftrightarrow \exists u_1\exists u_2\forall u'(u_1 < u'\ \&\ u' < u_2\ \&\ \mathrm{E}u' \leftrightarrow u' = u)$.

This is none too perspicuous, and psychologically it is doubtful that Cantor could have discovered even those of his results that *can* be thus reformulated without set language, had he not introduced set language at the stage he did. But *logically* the results prior to (3) *can* be reformulated without set language, both in words, as in (5), and in first-order logic, as in (6).

From a logical point of view, the existential generalization of (4), which tells us we really are getting a significantly stronger theorem when the hypothesis is weakened from (1) to (2), is especially interesting. It reads as follows:

(7) There is a set X of points such that $\partial X \neq \emptyset$ but $\partial^2 X = \emptyset$.

This, too, can be restated in terms of points rather than in terms of a point-set, along the lines of (5), as follows:

(8) There are some points such that not every point among them is isolated from the other points among them, but every point among them that is not isolated from all other points among them is isolated from all other points among them that are not isolated from all other points among them.

In this case, however, formalization in first-order logic, along the lines of (6), is impossible. The initial quantification, "there are some points . . ." is irreducibly plural, and cannot be rendered in terms of singular quantifiers "there is a point . . ." or "for every point . . ." The status of (7) is thus somewhere between that of (1) or (2) or (4) on the one hand, which can be given first-order formalizations quantifying only over points, and that of (3) on the other hand which cannot be stated without treating sets as objects.

The kind of irreducibly plural quantification exemplified by (8) was made an object of special study by the late George Boolos, who gave simpler examples of the phenomenon, and developed an extension of first-order logic to handle them. Analogous to (5) and its existential generalization (8) are the following:

(9) The critics who write for *Exceptional* magazine admire only each other.

(10) There are some critics who admire only each other.

Here (9) can be given a first-order formalization, while (10), attributed to Peter Geach and David Kaplan, cannot.

Irreducibly plural quantification amounts to the very last stop before the introduction of sets. If we are to understand what is most distinctive about set theory, we must understand what it adds to and how it goes beyond the mere plural; and to understand *that* we must first understand the plural and its logic.

2 PLURAL LOGIC

Unfortunately notation and terminology connected with plural quantification have not been standardized. Boolos (1984), Lewis (1991), and Burgess and Rosen (1997) all differ. The first item of business must be to indicate the notation and terminology to be used here. To begin with we have all the apparatus of singular logic. Thus we have $\sim$, &, $\vee$, $\rightarrow$, $\leftrightarrow$, $\forall$, $\exists$, and the logical predicate of identity $=$, along with appropriate logical axioms and rules of implication. In particular we have the usual laws of identity, namely, *self-identity* and *indiscernibility of identicals*:

(1) $u = u$

(2) $u = v \rightarrow (\Phi(u) \rightarrow \Phi(v))$.

Note the conventions used in displaying these laws. In both (1) and (2) initial universal quantifiers have been suppressed. In (2), what we really have is a *scheme* or rule according to which, for any formula $\Phi(t)$, writing $\Phi(u)$ and $\Phi(v)$ for the results of substituting u and of substituting v for each free occurrence of t therein, (2) is to count as an axiom. In (2) again, the technical proviso is left tacit that the variables u and v are free for t in $\Phi(t)$, which is to say that no free occurrence of t in $\Phi(t)$ occurs within the scope of a quantifier $\forall u$ or $\exists u$ or $\forall v$ or $\exists v$. In (2) yet again, it is to be understood that there may be *parameters*, or free variables not displayed, so that what is really to count as an axiom is something like:

(3) $\forall w_1 \ldots \forall w_n \forall u \forall v\, (u = v \rightarrow (\Phi(u, w_1 \ldots, w_n) \rightarrow \Phi(v, w_1 \ldots, w_n)))$

The derivability of the law of symmetry

(4) $u = v \rightarrow v = u$

depends on allowing parameters, the formula $\Phi(t)$ to which (2) is applied being $t = u$.

In terms of identity one may define certain other notions, notably *distinctness* and *unique existence*:

(5) $u \neq v \leftrightarrow \sim u = v$
(6) $\exists! x\Phi(u) \leftrightarrow \exists u\Phi(u)\ \&\ \sim\exists u_1 \exists u_2(\Phi(u_1)\ \&\ \Phi(u_2)\ \&\ u_1 \neq u_2)$.

Here the distinctness predicate $\neq$ and the unique existence quantifier $\exists!$ are to be treated not as *primitive* but as *defined*, which is to say that they are not part of the official notation, but rather are unofficial abbreviations. The definitions (5) and (6) do not count as substantive assumptions or axioms, but merely as abbreviations for tautologies, for biconditionals of the form $A \leftrightarrow A$.

So much for singular or first-order logic. The principles concerning the formulation of axioms and the status of definitions that have been set out at some length above in connection with identity are to be tacitly understood as still applying as we now move on to plural logic, and when we later move on to set theory. Turning then to plural logic, to begin with we need to add three items of notation. First, there are plural variables, *xx*, *yy*, *zz*, and so on. Second, there are plural quantifiers, written $\exists\exists$ and $\forall\forall$. If $\exists u$ and $\forall u$ are read "there is an object, *u*" and "for any object, *u*," then $\exists\exists xx$ and $\forall\forall xx$ may be read "there are some objects, the *x*s" and "for any objects, the *x*s." Third, there is a logical predicate of two places, with singular variables going in the first place and plural variables in the second, $u \propto xx$, which may be read "*u* is one of the *x*s" or "*u* is among the *x*s." Much as the symbol used in set theory for "element" is a stylized epsilon, the symbol used here for "among" is a stylized alpha.

A question immediately arises about the understanding of the plural quantifier. In a language with a threefold distinction among singular, dual, and plural, it would be natural to take "there are some objects . . ." to mean "there are three or more objects . . ." In a language like English, where we have only the distinction between singular and plural, it is natural to take it to mean "there are two or more objects . . ." For instance, in the Geach–Kaplan example (1.10) – that is, displayed item (10) of §1 above – it is natural to understand "some critics" as meaning two or more critics. But Lewis (1991) takes "there are some objects . . ." to mean "there are one or more objects . . .," while Burgess and Rosen (1997) go further and take it to mean "there are zero or more objects . . ." The last-named authors, however, show that taking any one reading as official, the other readings can be defined in terms of it, so in one important sense it does not matter which reading we take. The "zero or more" reading will be adopted here.

To frame a basic deductive system for plural quantification, one adds to complete such a system for singular quantification the following:

(7) *Axiom of Comprehension*
$\exists\exists xx \forall u(u \propto xx \leftrightarrow \Phi(u))$

(8) *Axiom of Indiscernibility*
$\forall u(u \propto xx \leftrightarrow u \propto yy) \rightarrow (\Phi(xx) \rightarrow \Phi(yy))$.

Axiom (7) is a plural analogue of existential generalization, allowing us, for instance, to infer (1.8) from (1.5). It says that for any condition there are some objects such that an object is among them if and only if the condition holds of it. Which objects are these? The objects of which the condition holds, of course! (If we adopted the "one or more" reading, comprehension would have to be formulated as the conditional with antecedent $\exists u\Phi(u)$ and consequent (7).) Axiom (8) is a plural analogue of the indiscernibility of identicals (4). It says that if exactly the same objects are among these objects as are among those objects, so that these and those are the very same objects, then any condition that holds of these holds of those.

One can introduce an additional notion and notation, not as primitive, but as defined, namely, "the *xs* are the same as the *ys*," the definition being as follows:

(9) $xx == yy \leftrightarrow \forall u(u \propto xx \leftrightarrow u \propto yy)$.

Then (8) can be abbreviated in the following form, to make clear the analogy with (2):

(10) $xx == yy \rightarrow (\Phi(xx) \rightarrow \Phi(yy))$.

No claim is made that the addition of the basic axioms (7) and (8) to a complete system of first-order logic produces a complete system of plural logic. For the moment, however, we have all the logical axioms and rules we will be needing.

3 EXTENSIONALITY

Two additional, non-logical primitives will be introduced in order to allow us to express the basic notions of set theory: ßu will be used to express "u is a set," and $u \equiv xx$ to express "u is the set of (all and only) the *xs*." (The symbol ß seems appropriate since as a ligature it is itself a one made out of a many, or at least out of a two.) Thus the primitive predicates of the *language of plural set theory*, to be used in an axiomatization here, are those of plural logic, $==$ and $\propto$, together with ß and $\equiv$; and the formulas of this language are built up from atomic formulas involving these predicates by means of $\sim$, &, $\vee$, $\rightarrow$, $\leftrightarrow$ and the singular and plural quantifiers $\forall$, $\exists$, $\forall\!\!\forall$, $\exists\!\exists$. By contrast, the primitive predicates of the *language of singular set theory* are that of singular logic, $=$, together with $\in$ and ß; and the formulas of this language are built up from atomic formulas involving these predicates by means of $\sim$, &, $\vee$, $\rightarrow$, $\leftrightarrow$ and the singular quantifiers $\forall$, $\exists$ only.

The first axiom to be assumed reads as follows:

(1) *Axiom of Heredity*
ß$u \leftrightarrow \exists xx(u \equiv xx)$.

The right-to-left direction of (1) expresses only the triviality that if an object is the set of some objects, then it is a set. The left-to-right direction would seem to express only the equally trivial converse, that if an object is a set, then it is the set of some objects; and this indeed is all it expresses if it is understood that the range of the quantifiers or universe of discourse includes all objects. However, what a sentence expresses changes if the universe of discourse is restricted, and if it is understood that the range of the quantifiers may be something less than all objects, what the left-to-right direction of (1) also expresses is that whenever a set is included in the universe of discourse, its elements are to be included along with it. It follows that if any of these elements are sets, their elements also are included, and if any of those elements are sets, their elements are included as well, and so on for generation after generation. Hence the name proposed for the axiom.

The second axiom involves a biconditional, and expresses in the two directions of the double arrow the features that distinguish the concept of a set made of elements from, on the one hand, the concept of a *whole* made of *parts* in *mereology*, and on the other hand, the concept of a *property* instantiated by *objects* in theories of *universals*. The axiom reads as follows:

(2) *Axiom of Extensionality*
$u \equiv xx \;\&\; v \equiv yy \rightarrow (u = v \leftrightarrow xx == yy)$.

The two features of the concept of set this axiom expresses are simply that, given a set, it is uniquely determined what elements it comprises, while given some elements, it is uniquely determined what set they compose. By contrast, a whole may be decomposable into parts in several different ways, while two properties may be instantiated by exactly the same objects and yet be distinct. While extensionality may not be an utter triviality, it is still in a sense less a substantive assumption than a partial explication of the *concept* of set. It would be inappropriate to use the word "set," rather than "whole" or "property" or whatever, for a concept of which extensionality was not a feature.

The relation of element to set, symbolized $\in$, which in conventional axiomatizations is taken as primitive, may now be introduced instead as an abbreviation. Either of the following will do as a definition:

(3) $v \in u \leftrightarrow ßu \;\&\; \exists xx(u \equiv xx \;\&\; v \propto xx)$
(4) $v \in u \leftrightarrow ßu \;\&\; \forall xx(u \equiv xx \rightarrow v \propto xx)$.

The equivalence of the right sides of (3) and (4) follows from (1) and the left-to-right direction of (2). It is also useful to introduce the plural version of $\in$, symbolized $\in\in$, by the following definition:

(5) $yy \in\in u \leftrightarrow \forall v(v \propto yy \rightarrow v \in u)$.

In terms of the defined notion $\in$, the right-to-left direction of (2) yields the following:

(6) $ßu \;\&\; ßv \rightarrow (\forall w(w \in u \leftrightarrow w \in v) \rightarrow u = v)$.

The last conditional could be strengthened to a biconditional, since the converse implication is immediate from the indiscernibility of identicals.

In conventional axiomatizations it is (6) or some variant version thereof that is usually called the axiom of extensionality. The feature that the elements determine the set is explicit in this formulation, while the intention to include within the range of the quantifiers or universe of discourse all the elements of any set that is itself, though it must be assumed since otherwise distinct sets might have the same elements within the restricted range or universe, is left implicit. The introduction of plurals, and the definition of $\in$ in terms of $\equiv$ have, so to speak, permitted what lies beneath extensionality in its conventional formulation to be exhibited explicitly in the form of the axioms (1) and (2). Making the assumption of heredity explicit will prove to be surprisingly important in §7 below.

Before leaving the topic of extensionality, it should be mentioned that the formulation of extensionality most commonly used in conventional axiomatizations is not (6) itself but a variant. Discussion of this issue requires some machinery we do not yet have, so it will be necessary to return to the issue of extensionality later on. For the moment we are done with this axiom.

The remaining axioms will be more substantive and will be existence assumptions. Before indicating what they are, let me recall that one tempting existence assumption *cannot* be made. One cannot assume that for *any* objects there is a set of just *those* objects:

(7) $\exists u(u \equiv xx)$.

For comprehension tells us the following:

(8) $\exists\exists xx \forall v(v \propto xx \leftrightarrow \sim v \in v)$.

Then (7) and (8) imply the following:

(9) $\exists u \forall v(v \in u \leftrightarrow \sim v \in v)$.

But (9) is a logical contradiction. All this is just the Russell paradox of the set of all sets that are not elements of themselves, adapted to the context of plural quantification.

This may be the best place to interject a word on the role of alleged entities called "classes" in the axiomatization of set theory. The word "class" rather than "set" was originally used by Frege for the extensions of concepts, but it has since come to be used for set-like entities that in some mysterious way fail to be sets. Usually though not invariably classes are spoken of as having "members," while sets have "elements." Thus membership is an elementhood-like relationship that in some mysterious way fails to be elementhood. There is a mystifying distinction between two fundamentally different kinds of collections, sets and classes, and two fundamentally different kinds of belonging, elementhood and membership.

Such, at any rate, is the situation described in the language of *splitters*, for whom no set is a class, though every set is *coextensive* with some class in the sense that the elements of the set are all and only the members of the class. On a different usage, favored by *lumpers*, one has not just coextensiveness but outright identity. Sets simply are some but not all of the classes, the others being called *proper* classes; the sets are usually distinguished from the proper classes in that each set is assumed to be a member of some class, while no proper class is a member of any proper class. The mysterious distinction between two fundamentally different kinds of collections, sets and classes, is replaced by the mysterious distinction between two fundamentally different kinds of classes: those that can and those that cannot be members of other classes.

If one is willing to accept the mystery, then classes have certain uses in set theory. In conventional axiomatizations, they permit the reduction of certain schemes (separation and replacement) to single axioms. But as Boolos urged, in this use plural quantification over sets can be employed to replace singular quantification over classes, as will be done in the next section. Classes are also appealed to in heuristic motivations for certain extensions of the conventional axioms, so-called *large cardinal* axioms, and the fact that they are thus appealed to has been argued by Schindler (1994) and others to weaken the case for large cardinals. Recently Uzquiano (2003) has pointed out that at least in most cases of such use, plural quantification over sets can again be employed to replace singular quantification over classes, as will be done in one important case in the section after next. The advantage of employing plural quantification in this way is that it leaves us able to maintain that there is just *one* kind of collection, and that set theory is the *most general* theory of collections.

It might be thought that, conversely, anything that can be accomplished with plurals can be accomplished with classes, enabling us to avoid appeal to an extra-classical logic. But plural quantification over sets *cannot*, in fact, always be replaced by singular quantification over classes, a point emphasized by Boolos and worth reemphasizing here. To begin by restating in words what has already been stated in symbols in (8) and (9) above:

(10) It is true that there are some objects such that all and only those objects that are sets and not elements of themselves are among them.

(11) It is false that there is a set of objects such that all and only those objects that are sets and not elements of themselves are elements of it.

So "there are some objects . . ." cannot in general be replaced by "there is a set of objects . . ." Class theory indeed tells us that

(12) There is a class of objects such that all and only those objects that are sets and not elements of themselves are members of it.

But still "there are some objects" cannot *in general* be replaced by "there is a class of objects . . .," and one need only replace each occurrence of "set" or "element" in (10) and (11) by "class" or "member," respectively, to see why not.

Another limitation of class-and-set theory is that, unlike plural-and-singular set theory, it is in a sense incapable of making explicit the assumption of heredity underlying extensionality. Technically, it would be possible to replace the plural-logical (3.1), asserting that every set is the set of some elements, by a class-theoretic version, asserting that every set is coextensive with some class. But to state this last while keeping elementhood primitive, and coextensiveness as defined, would be merely to restate one specific instance of the general scheme of comprehension for classes. For a genuine statement of the heredity assumption, one would need to take the notion of a set's being coextensive with a class to be primitive, and the notion of an object's being an element of a set to be defined in terms of it, namely, defined as the object's being a member of the class with which the set is coextensive. And that choice of primitives is extremely unnatural from a class-theoretic point of view, and to my knowledge no class theorist has ever made that choice. Of course, the importance of this limitation cannot be clear until it is seen what use is made of the heredity assumption in the further development of set theory.

4 LIMITATION OF SIZE

Let us return to that development. The name "extensionality" is a reminder of Frege's notion of the "extension of a concept." Frege really did assume that every condition determines a set, or more precisely, determines a "concept" and thereby an "extension," and so fell into paradox. Even before the discovery of any such paradoxes, however, Cantor (1885) had rejected Frege's assumptions in a review of Frege's *Grundlagen*:

> The author's own attempt to give a strict foundation to the number-concept seems to me less successful. Specifically, the author has the unfortunate idea . . . to take as the foundation for the number-concept what in Scholastic logic is called the "extension of a concept." He completely overlooks the fact that in general the "extension of a concept" is quantitatively wholly indeterminate. Only in certain cases is the extension quantitatively determinate, in which cases it can then of course be assigned a definite number, if it is finite, or power, in case it is infinite. But for this sort of quantitative determination we must already possess the concepts of "number" and "power," and it is *getting things backwards* to try to found these latter concepts on the concept "extension of a concept."

Exactly what Cantor meant by "quantitatively indeterminate" (*quantitativ unbestimmt*) is not entirely clear, but he seems to be alluding to the kind of distinction within the "actually infinite" that he makes elsewhere, between merely "transfinite" or "consistent multiplicities," which do form sets, and the "absolutely infinite" or "inconsistent multiplicities," which do not. The Russell paradox does not arise for Cantor because he never assumes that every condition determines a set, but only those conditions that do not hold of *too many* objects, an assumption known as the principle of *limitation of size*.

The principle is subject to differing interpretations, based on differing understandings of what it is to be "too many" to form a set. In advance of any deep analysis of how to measure "size," however, it is already clear that the limitation of size principle motivates one important axiom, namely, Zermelo's axiom of *separation*, according to which if given things are not too many to form a set, then any things among them are also not too many to form a set; or to put the matter another way, given any things that are among the elements of some set, they also may be "separated out" from the other elements of the set to form a subset.

Formally, one way to state the axiom is as follows:

(1) $ßu \rightarrow \exists v(ßv \mathbin{\&} \forall w(w \in v \leftrightarrow w \in u \mathbin{\&} w \propto zz))$.

It is not hard to see that (1) is equivalent to the following:

(2) $ßu \mathbin{\&} \forall w(w \propto yy \rightarrow w \in u) \rightarrow \exists v(v \equiv yy)$.

And to eliminate the use of the defined symbol $\in$, it is also not hard to see that (2) is equivalent to the following:

(3) *Axiom of Separation*
$\forall w(w \propto yy \rightarrow w \propto xx) \rightarrow \exists u(u \equiv xx) \rightarrow \exists v(v \equiv yy))$.

Separation could equivalently be formulated as a scheme:

(4) $ßu \rightarrow \exists v(ßv \mathbin{\&} \forall w(w \in v \leftrightarrow w \in u \mathbin{\&} \Phi(w)))$.

For (1) is implied by (4), being simply the instance with $w \propto zz$ as $\Phi(w)$; while conversely comprehension, which for any $\Phi(w)$ gives us *zs* such that $w \propto zz$ if and only if $\Phi(w)$, together with (1) implies (4). But clearly no one becomes convinced of a scheme by becoming convinced separately of each of its instances, one by one; there must be some single underlying principle, and much as (3.1) made explicit an assumption underlying conventional formulations of extensionality, so (3) above makes explicit the single assumption underlying schematic formulations of separation. Hence (3) will be taken as our official version. Formulating separation as a single axiom rather than a scheme will prove to be important in §7 below.

However formulated, the assumption of separation is so fundamental to Cantorian thought that it is arguably inappropriate to apply Cantor's word "set" (*Menge*) to theories (such as Quine's NF and ML) that do not accept it. In other words, separation may be regarded as a partial explication of the concept of set, indicating what sets are supposed to be like if they exist. It is also, however, a kind of existence assumption. Specifically, it is a *relative* existence assumption that, given one set, provides us with others. But it is not a *positive* existence assumption. The axioms adopted so far, taken together, do not yet imply the existence of a single set.

Now as has been said, on virtually any understanding of "many" and related notions, separation follows from the principle that objects form a set unless they are too many, because if all of *these* are among *those*, then there are no more of these than of those. Likewise, if there are just as many of *these* as of *those*, and if those are not too many, then these are not too many either. To extract an axiom from this latter thought, however, we need to adopt some specific understanding of "just as many."

But here Cantor's analysis in terms of a *correspondence* between these and those is the only live option. Understanding "just as many" according to this analysis, we get Frankel's axiom of *replacement*, according to which if some things form a set, and some other things are in correspondence with them, then those other things also form a set; or to put the matter another way, if to each element of a set there corresponds some one other object, then the elements of the set may be "replaced by" the corresponding objects, and we will still have a set. The more precise and formal statement need not detain us here.

To obtain the further axioms of the usual Zermelo–Frankel (ZF) set theory, one needs further specific judgments or interpretations of how many objects are "too many" to form a set. Assuming that two is not too many, we get the axiom of *pairing*, assuming – contrary to tradition, but following Cantor – that infinitely many is not too many, we get the axiom of *infinity*, and so on. Yet further judgments or interpretations of this kind suggest axioms beyond those of ZF, the large cardinal or higher infinity axioms alluded to earlier, in an open-ended series of increasing strength. But at present such large cardinal assumptions, not included in ZF, generally are not taken for granted as axioms by mathematicians. If a theorem depends on such an assumption, it must be stated as part of the hypothesis of the theorem.

By contrast, the famous or notorious axiom of choice (AC) generally is no longer, as it once was, stated as part of the hypothesis of a theorem that depends on it. Today it is generally taken for granted as an axiom, and the usual system of axiomatic set theory is ZF + AC = ZFC. The main discussion, and even the statement, of AC will be postponed to a later section, but it may be mentioned here that von Neumann has argued that even AC can be motivated as an additional axiom by appeal to a certain interpretation of the principle of limitation of size. According to this interpretation, called the *maximality principle*, given objects form a set unless *there are as many of them as there are objects altogether*. How AC follows from this principle is among the many points explained in the thorough study of the history of the principle of limitation of size from Cantor through Zermelo to von Neumann and beyond undertaken in Hallett (1984).

In the light of this study it can be said that the usual axioms of set theory, and large cardinal axioms also, can be motivated by appeal to a single principle, that of limitation of size, provided we allow ourselves an open-ended series of increasingly more specific understandings of that principle. But what we do *not* yet have is a *single* axiom formalizing a *single* understanding of the principle of limitation of size, from which the usual axioms can be formally deduced. If we are to obtain set theory from a single positive existence axiom in addition to those already adopted, another approach must be tried.

5 THE REFLECTION PRINCIPLE

Another approach is available from the work of Paul Bernays (1961), building on the work of Azriel Levy, and this rival approach seems more promising than an appeal to the maximality principle. The background is as follows. Levy derived from the usual axioms of set theory a result known as the *reflection principle*, whose precise statement need not detain us. Bernays showed that most of the usual axioms could be deduced from a strengthened version of the principle, which thus could be adopted as an axiom in their place. He also showed that several large cardinal axioms – for the *cognoscenti*, those known as the axioms of *inaccessible* and of *Mahlo* cardinals of all orders, an upper bound for the cardinals obtainable in this way being *indescribable* cardinals – could also be deduced from his version of the reflection principle.

One may attempt to motivate the principle by appeal to the idea of limitation of size. This heuristic motivation is perhaps best presented as a list of principles, the first of which is Cantor's vague principle of limitation of size, each succeeding one of which represents a plausible way of making the preceding one more precise, and the last of which will provide the basis for the formal axiom whose consequences are to be explored. Without further ado, here is the list. All principles have the general form: "The *xs* form a set unless . . ."

(1) . . . they are indeterminately or indefinitely many.
(2) . . . they are indefinably or indescribably many.
(3) . . . any statement Φ that holds of them fails to describe how many they are.
(4) . . . any statement Φ that holds of them continues to hold if reinterpreted to be not about all of them but just about some of them, fewer than all of them.
(5) . . . any statement Φ that holds of them continues to hold if reinterpreted to be not about all of them but just about some of them, few enough to form a set.

(Note that the transition from (4) to (5) would be automatic assuming the maximality principle.)

Our interest will in fact be limited to the case where the *xs* are *all* objects. If our general understanding is that our quantifiers range over all objects, then *every* statement is about all objects, and the principle thus becomes:

(6) Any statement Φ that holds continues to hold if reinterpreted to be just about the elements of some set t.

With respect to what is expressed by Φ, the macrocosm of all objects is "reflected" in the microcosm of elements of t.

One must be careful about how one applies this reflection principle. One thing that holds is that *no object is a set that has all objects as elements.* Applying reflection might seem to give a set t for which it holds that *no object is a set that has all elements of t as elements*; and, of course, there can be no such set. This example, however, is an improper application. Proper application only yields a set t for which it holds that *no element of t is a set that has all elements of t as elements*; and this is entirely non-paradoxical. But clearly what is needed at this point is a formal restatement.

As a preliminary we need the notion of *relativizing* or restricting quantifiers to a formula $\Theta(u)$. By this we mean replacing statements about some or any object or objects by statements about some or any object or objects of which the predicate holds. Formally, given any formula Φ, we form the relativization Φ^{Θ} by replacing each quantification of the kind shown on the left side below by the quantification shown beside it on the right side.

$\forall u$	$\forall u(\Theta(u) \rightarrow \ldots)$
$\forall\forall xx$	$\forall xx(\forall u(u \propto xx \rightarrow \Theta(u)) \rightarrow \ldots)$
$\exists u$	$\exists u(\Theta(u) \;\&\; \ldots)$
$\exists\exists xx$	$\exists\exists xx(\forall u(u \propto xx \rightarrow \Theta(u)) \;\&\; \ldots)$

Of special interest will be the case where the formula $\Theta(u)$ is $u \in t$ for some parameter t. In this case one writes Φ^t for Φ^{Θ} and calls it the relativization to t.

In this connection it is often useful to use the abbreviations on the left side below for the expressions on the right side:

$\forall u \in t\,(\ldots)$	$\forall u(u \in t \rightarrow \ldots)$
$\forall\forall xx \in\in t\,(\ldots)$	$\forall xx(xx \in\in t \rightarrow \ldots)$
$\exists u \in t\,(\ldots)$	$\exists u(u \in t \;\&\; \ldots)$
$\exists\exists xx \in\in t\,(\ldots)$	$\exists\exists xx(xx \in\in t \;\&\; \ldots)$

Then the replacements needed to obtain Φ^t amount to the following:

$\forall u$	$\forall u \in t\,(\ldots)$
$\forall\forall xx$	$\forall\forall xx \in\in t\,(\ldots)$
$\exists u$	$\exists u \in t\,(\ldots)$
$\exists\exists xx$	$\exists\exists xx \in\in t\,(\ldots)$

Our principle then becomes the following *reflection* axiom:

(7) *Axiom of Reflection*:
$\Phi \rightarrow \exists t \Phi^t$.

Note that (7) is meant to apply to formulas written out in primitive notation, without abbreviations, or at worst, involving only abbreviations that, like $\neq$ and unlike $\exists!$, do not involve any quantifiers in their definitions. An abbreviation involving quantifiers needs to be written out to make the quantifiers explicit, so that they can be relativized.

6 DEDUCING EXISTENCE AXIOMS OF ZF

The set theory based on plural logic and the non-logical axioms of heredity (3.1), extensionality (3.2), separation (4.3), and reflection (5.7) will be called BB for Bernays–Boolos. It turns out that all the usual existence axioms of ZF, as well as the large cardinals obtainable in the manner of Bernays, can be obtained from BB by deduction as logical consequences.

Bernays was not especially concerned with the intuitive or heuristic motivation for reflection, and in fact assumed reflection in an ostensibly stronger version than (5.7), with an additional technical condition on t that it is not immediately obvious how to motivate by appeal to the idea of limitation of size. Our first task must be to show that this ostensibly stronger version actually follows from the version (5.7) that has been taken as axiomatic here.

The first step in the deduction is to replace the version of reflection in (5.7), which does not even explicitly state that t is a set, by a version that does state explicitly at least that much. The deduction uses a trick employed over and over by Bernays. The trick consists in noting that any statement Φ of course implies its own conjunction with any axiom or theorem Ψ. And so (5.7) yields the following:

(1) $\Phi \rightarrow \exists t(\Psi^t \,\&\, \Phi^t)$.

Taking as Ψ the trivial truism $\exists u(u = u)$ we get the following:

(2) $\Phi \rightarrow \exists t(\exists u(u \in t \,\&\, u = u) \,\&\, \Phi^t)$.

The existence of an element u of t implies that t is a set by (3.3), so we have the following:

(3) $\Phi \rightarrow \exists t(ßt \,\&\, \Phi^t)$.

The derivation of (3) is a trivial illustration of how (5.7) can yield stronger versions of reflection.

For a slightly less trivial instance, take as Ψ the axiom (3.1). What we get is the following:

(4) $\Phi \rightarrow \exists t(ßt \,\&\, \forall u \in t\ (ßu \leftrightarrow \exists\exists xx \in\in t(u \equiv xx)) \,\&\, \Phi^t)$.

What the left-to-right direction of the second conjunct $(3.1)^t$ says is that any set that is an element of t is the set of some objects that are themselves elements of t. This is merely a long-winded way of saying that t is a set such that every element of an element of t is an element of t, a property called the *transitivity* of t. Thus we have obtained the following strengthening of (5.7):

(5) $\Phi \rightarrow \exists t(t \text{ is transitive } \& \ \Phi^t)$.

It was, of course, crucial to the derivation of (5) that $\equiv$ has been taken as primitive. For recall that in relativizing a formula, abbreviations defined using quantifiers are supposed to be written out in primitive terms. In principle this includes even the symbol $\in$, which here is officially an abbreviation defined by (3.3). What $(v \in u)^t$ amounts to is the following:

(6) $\exists\exists xx \in\in t(u \equiv xx \ \& \ v \propto xx)$.

If, however, t is transitive, and u is an element of t, then it will automatically be the case that the objects of which u is the set will all be elements of t, and therefore (3) is equivalent to the simple $v \in u$. In other words, in practice the abbreviation $\in$ does *not* need to be written out in primitive terms after all, so long as the set t can be taken to be transitive, which by (5) it always can be.

The principle (5), with Φ allowed to contain $\in$, is the version of reflection assumed by Bernays, apart from his using the language of singular quantification over classes rather than that of plural quantification over sets, which makes no difference to the subsequent deduction of further existence axioms. Thus at this point one could simply cite Bernays for the deduction of further axioms, and close the present section.

It may be instructive, however, to indicate the first few steps in the further deduction. The first step will be yet another application of the trick that took us from (5.7) to (3) and from (3) to (5). A transitive set t is said to be *supertransitive* if any *subset* of any element of t, as well as any element of any element of t, is itself an element of t. By taking as Ψ in (1) the axiom (4.3), the following further strengthening is also obtainable:

(7) $\Phi \rightarrow \exists t(t \text{ is supertransitive } \& \ \Phi^t)$.

Note that it was crucial for obtaining supertransitivity that separation is formulated as a single axiom (4.3), not as a scheme (4.4). For the single-axiom formulation one needs either to use singular quantification over classes, as Bernays did, or plural quantification over sets, as was done here. If one attempts to stick with singular quantification over sets, it will be impossible to obtain supertransitivity, and in consequence impossible to obtain certain existence axioms.

But having strengthened (5) to (7), Bernays rapidly obtains the axioms of *pairing, union, power, infinity, replacement*. For instance, following a deduction he attributes to Klaus Gloede, given any two objects a and b, we take as our Φ the following logical truth:

(8) $\exists u(u = a) \,\&\, \exists u(u = b)$.

Reflection according to (3) then gives us the following:

(9) $\exists t(ßt \,\&\, \exists u \in t(u = a) \,\&\, \exists u \in t(u = b))$.

And this in turn implies the following:

(10) $\exists t(ßt \,\&\, a \in t \,\&\, b \in t)$.

This already is an alternative formulation of the axiom of pairing. A more usual formulation calls for the existence of a set t whose elements include a and b and *nothing else*; but this follows immediately from (10) on applying separation (4.4) to the condition $u = a \vee u = b$. The most usual formulation calls for the existence of a *unique* set, denoted $\{a, b\}$, whose elements include a and b and nothing else; but this follows immediately on applying extensionality (3.6).

The deductions of union and of power are almost equally easy and would make good exercises for the reader familiar with the statement of the axioms. As a hint it may be mentioned that for union, transitivity is needed; for power, supertransitivity. For replacement, and for infinity and "higher infinities" or large cardinals, for which last supertransitivity and hence the single-axiom formulation of separation is again needed, as it was for power, the reader is referred to Bernays.

7 EXTENSIONALITY REVISITED

To obtain full ZFC we need to obtain three more axioms. The first of these is a stronger version of extensionality that reads as follows:

(1) $\forall w(w \in u \leftrightarrow w \in v) \rightarrow u = v$.

This will be forthcoming from (3.6) provided we assume the following:

(2) *Axiom of Purity*
$ßu$.

From one point of view (2) is an utter absurdity, saying that there are no objects but sets. Now the mere absurdity of a proposition is no guarantee that some philosopher will not endorse it. So perhaps there is a philosopher somewhere who denies that anything but sets exists, and who denies that he himself exists. Or perhaps there is a philosopher somewhere who denies that anything but sets exists, and maintains that she herself is a set. But this is not what is usually intended by those who adopt (2). From the usual point of view, (2) merely expresses the intention to exclude anything but sets from the range of the quantifiers. Note, however, that since we are already assuming that whenever a set is included its elements are included as well, the assumption (2) actually involves a restriction not merely to sets, but to sets all of whose elements are sets, and all elements of whose elements are sets, and so on. Only *pure* sets are included, hence the name for (2).

Now there are two approaches that might be taken to obtain (2) and thence (1). One would be to take it as an axiom. Another would be to try to find an interpretation of set theory *with* the axiom (2) within set theory *without* the axiom (2). That is, one could try to find a formula Π such that each axiom of set theory *with* axiom (2) becomes a theorem of set theory *without* axiom (2) when quantifiers are relativized to Π. Then for every theorem Φ of set theory with axiom (2), one would have a theorem Φ^{Π} of set theory without axiom (2) saying that Φ holds *for those sets for which condition* Π holds.

Ideally, the formula $\Pi(u)$ should be one that intuitively expresses that u is a pure set, or in other words, that u, its elements, the elements of its elements, and so on, are all sets. Now intuitively the set u, its elements, the elements of its elements, and so on, are together some objects such that u is among them and any element of a set among them is among them. Thus intuitively if u is pure, there will be some objects such that any element of a set among them is among them, u is among them, and every object among them is a set. Conversely, if there are some objects such that any element of a set among them is among them, and if u is among them, and every object among them is a set, then it follows that u, the elements of u, the elements of elements of u, and so on, are all sets, and u is pure. Thus a natural candidate for the formula $\Pi(u)$ would be the following:

(3) $\exists\exists xx(\forall v \forall w(v \propto xx \mathbin{\&} w \in v \rightarrow w \propto xx) \mathbin{\&} u \propto xx \mathbin{\&} \forall v(v \propto xx \rightarrow \text{ß}v))$.

Reflection can be used to show (3) is equivalent to the following alternative not involving plural quantifiers:

(4) $\exists t(\forall v \forall w(v \in t \mathbin{\&} w \in v \rightarrow w \in t) \mathbin{\&} u \in t \mathbin{\&} \forall v \in t\, \text{ß}v)$.

The first conjunct of (4) says that t is transitive. Other variations in the choice of Π are also possible.

Whatever variant is chosen, what has to be proved is, first, that for each axiom Φ that is being assumed, one can deduce its relativization Φ^{Π} from the axioms that are being assumed, and, second, for the axiom (2) that is *not* being assumed, one can also deduce its relativization $(2)^{\Pi}$ from the axioms that are being assumed. It follows that whenever Φ follows from the axioms that are being assumed plus (2), then Φ^{Π} follows from the axioms that are being assumed.

Though the tedious verification of details will not be given here – for they are quite similar to the details involved in the case of the axiom of foundation, to be discussed below – this is in fact the case whether by "the axioms that are being assumed" one means the other axioms of ZFC or means the axioms of BB. Moreover, it is also true that for any large cardinal axiom, even if stronger than those provided by BB, that one might decide to assume in the future, the assumption of such an axiom Φ would render its relativization Φ^{Π} deducible as well. Thus one may say that, *without losing any other axiom that one might want*, axiom (2) and hence axiom (1) can be "obtained," not in the sense of being got by deduction as a logical consequence, but in the sense of being got by relativization or restriction of quantifiers.

An arguable advantage of the first approach of assuming purity as an axiom is that it permits a simpler axiomatization. One can introduce what amounts to a function symbol εε, which applied to a singular variable u yields a term $\varepsilon\varepsilon u$ of a kind that can be substituted for plural variables, and whose intended meaning is "the elements of u." The symbol ß can be dropped, and the symbol $\equiv$ need not be taken as primitive, but rather may be taken as defined by the following:

(5) $u \equiv xx \leftrightarrow \varepsilon\varepsilon u == xx.$

In general, the use of function symbols builds existence and uniqueness assumptions into the notation, making it unnecessary to assume them as axioms. In the present case, with εε there is no need for the axiom of heredity, and no need for one direction of the axiom of extensionality. All that needs to be assumed as an axiom of extensionality is the following:

(6) $\varepsilon\varepsilon u == \varepsilon\varepsilon v \rightarrow u = v.$

The converse follows from the indiscernibility of identicals.

By contrast, the advantage of the second approach of restricting quantifiers is the philosophical one that it makes explicit what the first approach leaves implicit, namely, that purity is not a substantive assumption, but a restriction on the universe of discourse. It is the second approach that will be adopted officially here. Thus the ZFC version of the extensionality axiom will *not* be added to the axioms of BB, and any theorems of ZFC that depend on it will not be theorems of BB; but for any such theorem Φ, the relativization Φ^{Π} *will* be a theorem of BB, or in other words, it will be a theorem of BB that Φ holds of *pure* sets.

The second of the three axioms of ZFC requiring discussion here is the axiom of *foundation*, also known as *regularity*. As axiom (2) implies that all objects (in the range of the quantifiers) are pure sets, so foundation or regularity asserts that all objects (in the range of the quantifiers) are *well-founded* sets. Since all elements of sets included in the universe of discourse are themselves included, it follows that if a set u is included, not only is u well-founded itself, but so are the elements of u, the elements of elements of u and so on: u is *hereditarily* well founded.

Several equivalent formal versions of foundation are known, but it will not be necessary to give *any* formulation here, since the question of the status of this axiom is already well explained – with details of the kind omitted in the discussion of purity above – in introductory textbooks, such as Hrbacek and Jech (1999). The options are the same as in the case of purity. Foundation may either be taken as an axiom, while insisting that it is not a substantive assumption, but merely an expression of the intention to limit what is included in the range of the quantifiers; or it may be obtained from the other axioms by relativization. The latter approach has the advantage of making explicit what the former leaves implicit, and will be adopted officially here.

8 THE AXIOM OF CHOICE

We now have "obtained," in one way or another, or one sense or another, all the axioms of ZF, and it "only" remains to consider AC. Having set aside von Neumann's approach based on his maximality interpretation of the principle of limitation of size, two approaches remain, and one is exactly similar to the approach to purity and foundation just taken. That is to say, choice can be obtained from the other axioms, not by deduction as a logical consequence, but by relativization or restriction of quantifiers.

For Gödel's famous proof of the relative consistency of choice – his proof that if ZF is consistent, then ZFC is consistent – proceeds by defining a technical notion of *constructible* set, and proving in ZF that all the ZF axioms, and in addition AC, hold when all quantifications are replaced by quantifications restricted to constructible sets. If large cardinal axioms in the range obtained by Bernays are considered, they also continue to hold when all quantifications are thus restricted, and this is true even for some larger large cardinals, though for still larger large cardinals one needs to look for some modification of Gödel's method, which for all but the largest large cardinals has indeed been found. It is, moreover, tedious but routine to adapt this approach to an axiomatization based on plural logic.

At a philosophical level, however, there is a great difference between the case of choice and that of foundation or purity. For it emphatically cannot be claimed that when Zermelo originally affirmed the choice axiom, or when Gödel himself and most later set theorists reaffirmed that axiom, they merely intended to exclude non-constructible sets from the universe of discourse. On the contrary, Zermelo was unacquainted with the sophisticated notion of the constructible set, while Gödel and many later set theorists have explicitly denied that all sets are constructible. Here one faces a decision of principle. If the goal is to provide an intuitive motivation for the axioms, the acceptability of relativizing quantifiers will be limited to cases like purity and foundation, where the relativization arguably represents the intentions of those who adopt the axiom in question.

By contrast, if the interest lies elsewhere, in finding a *reinterpretation* of set theory under which all its axioms will be derivable from a *minimal* basis, the method of relativizing quantifiers may be usable without limitation. It should be noted, however, that if one does allow reinterpretation without limitation, then it is possible to do very much better in the way of obtaining set theory from a minimal basis than has been indicated so far. For recent work of Harvey Friedman shows that without the apparatus of plural logic, without any axiom of heredity, without any version of an axiom of extensionality, and without a separate axiom of separation, *all* of set theory can be reinterpreted in a system whose *sole* axiom is a suitably formulated reflection principle. Moreover, the formulation of the principle can be modulated so as to obtain, as one wishes, ZFC without the power axiom, or ZFC exactly, or ZFC with the kind of large cardinals obtained by Bernays, or ZFC with much larger large cardinals. But since our interest here has been in the more direct motivation of axioms, Friedman's impressive results will be left aside.[1] Along with them will be left aside Gödel's approach to obtaining AC using his constructible sets.

Having set aside other approaches, one last approach to motivating the axiom of choice remains to be considered. Or rather, I should say, one last

[1] Friedman makes his papers available, between the time of their writing and the time of their publication, on the preprint server: www.mathpreprints.com/math/Preprint/show. Type "Harvey Friedman" into the window for access to these preprints.

approach to *obtaining* the axiom, leaving it to the reader to judge how far this approach motivates the axiom. But to begin with, the statement of the axiom should be given. It reads as follows:

(1) $\forall u(u \in a \rightarrow \exists w(w \in u)\ \&\ \forall u \forall v(u \in a\ \&\ v \in a \rightarrow \forall w(w \in u \leftrightarrow w \in v) \vee {\sim}\exists w(w \in u\ \&\ w \in v)) \rightarrow \exists v \forall u(u \in a \rightarrow \exists! w(w \in u\ \&\ u \in v))$.

If we call the sets in *a distinguished*, then the first clause of the hypothesis says that distinguished sets are non-empty, and the second that distinguished sets are non-overlapping: a distinguished u and v either have exactly the same elements (and hence by extensionality are exactly the same set) or else have no common elements at all. The conclusion asserts the existence of what is called a *choice* set, having exactly one element in common with each distinguished set.

Now even before set-theoretic notions are introduced, there is a version of the axiom of choice that can be formulated as a purely *logical* assumption. It is a scheme, reading as follows:

(2) $\forall\forall xx(\Phi(xx) \rightarrow \exists w(w \propto xx))\ \&\ \forall\forall xx \forall\forall yy(\Phi(xx)\ \&\ \Phi(yy) \rightarrow \forall w(w \propto xx \leftrightarrow w \propto yy) \vee {\sim}\exists w(w \propto xx\ \&\ w \propto yy)) \rightarrow \exists\exists yy \forall\forall xx(\Phi(xx) \rightarrow \exists! w(w \propto xx\ \&\ w \propto yy))$.

If we call objects distinctively related to each other, or *distinguished* for short, when condition Φ holds of them, the two conjuncts of the hypothesis of the axiom are non-emptiness and non-overlappingness conditions for distinguished objects. The conclusion asserts the existence of some objects, which may be called the *chosen* objects, such that for any distinguished objects, there is exactly one chosen object among them.

It is an easy exercise to deduce (1) from (2) in our framework, taking as $\Phi(xx)$ in the condition:

(3) $\exists u \in a\ (u \equiv xx)$.

The hypothesis of (1) easily gives us the hypothesis of (2) for (3) as Φ. What the conclusion of (2) for (3) as Φ gives us is some chosen objects, the *ys*, while what the conclusion of (1) demands is a choice set. What is left to the reader is to verify that the *ys* do form a set v as required.

It is not quite so easy an exercise to show, though it is also true, that given the set theory BB, the set-theoretic version of AC in the form (1) implies (each instance of) the logical version of AC in form (2). (This derivation of (2) from (1) is an instance of a more general phenomenon of the derivability of logical conclusions from set-theoretic assumptions, to be discussed in the next section.) What one proves is the contrapositive, that if (2) fails (in a particular instance), then (1) fails. From the assumed failure of (2) for a particular Φ, one obtains by reflection a supertransitive set t for which $\sim(2)^t$ holds and hence $(2)^t$ fails. One can then take as a the set of all subsets u of t such that

(4) $\exists\exists xx(u \equiv xx \ \& \ \Phi_t(xx))$.

The non-emptiness and non-overlap clauses in the antecedent of $(2)^t$ will then imply the corresponding clauses in the antecedent of (1) for the set a defined by (4). But if there were, as per the consequent of (1), a choice set v for a, then its elements, call them the *ys*, would be chosen elements, as per the consequent of $(2)^t$, contrary to the hypothesis that $(2)^t$ fails.

It follows that one can obtain the usual set-theoretic version of AC *without* adding any new set-theoretic axioms to BB, by adding a version of AC to the background plural logic. This is the course that will be adopted here, but the logical version of AC adopted will not be (2) above, but something else, whose introduction requires some background.

The most important points about AC in the present context would be the following. On the one hand, AC is useful for proving many mathematical results in their most general form. This pragmatic consideration has historically and practically been the most widely *persuasive* motivation for the axiom. It is for this reason that, as stated earlier, mathematicians have generally acquiesced in its assumption: it is the reason why today only logicians still keep track of which theorems require AC and which do not.

On the other hand, AC is a *non-constructive existence assertion*, asserting that something exists for which a given condition holds, without specifying any particular such thing. That is why when it was first introduced mathematicians did not at once embrace it. But AC is not the only or the most basic non-constructive existence assertion in classical mathematics or logic, since the following basic law of monadic first-order logic is also such an assertion:

(5) $\exists u(\exists v\Phi(v) \rightarrow \Phi(u))$.

That is why it is now widely agreed that if one is going to object to non-constructive existence assertions, one should not wait for AC, but should begin objecting already at the level of classical logic, if not of classical sentential logic, thus placing one's objections entirely outside the scope of the present paper. And inversely, those who are determined, in the words of Hilbert, not to be driven out of "Cantor's paradise," must begin their defensive operations no later than the level of monadic first-order logic, or indeed sentential logic – "Boole's paradise," as it might be called.

Hilbert, and with him Bernays, starting from this last observation, proposed a way of building AC even into first-order, singular logic, where it cannot otherwise be expressed, even as a scheme after the pattern of (2). They allow the formation of a term $\varepsilon u[\Phi(u)]$ which can be substituted for free variables. It may be thought of as a description, "the chosen u such that $\Phi(u)$," provided that this is understood in such a way that, when there is in fact no u such that $\Phi(u)$, what the phrase denotes is some arbitrarily chosen object.

Accordingly the following logical axiom for the ε-symbol is assumed:

(6) $\exists u \Phi(u) \rightarrow \Phi(\varepsilon u[\Phi(u)])$.

(Actually, a biconditional version of (6) was for Hilbert the very definition of the existential quantifier, though few have followed him in this, and we will not.) The connection with (5) is apparent.

The following is also sometimes assumed:

(7) $\forall u(\Phi(u) \leftrightarrow \Psi(u)) \rightarrow \varepsilon u[\Phi(u)] = \varepsilon u[\Psi(u)]$.

This says that what the chosen object is depends not on the condition Φ but on what things it holds of, so that if Ψ holds of exactly the same things, the same object will be chosen.

If the ε-symbol is added to *plural* logic, then in terms of it we can define the chosen object among some given objects:

(8) $\alpha xx = \varepsilon u[u \propto xx]$.

(If we had the $\varepsilon\varepsilon$ notation of the preceding section, $\alpha\varepsilon\varepsilon u$ would amount to "the chosen element of u" if u is non-empty.) But actually, in this case it is more natural to take α as the primitive notion, and let ε be defined by:

(9) $v = \varepsilon u[\Phi(u)] \leftrightarrow \exists\exists xx(\forall u(u \propto xx \leftrightarrow \Phi(u))\ \&\ v = \alpha xx)$.

To derive (6) it is enough to assume the following:

(10) *Axiom of Choice*
$\exists u(u \propto xx) \rightarrow \alpha xx \propto xx$.

It is this version that will be taken officially as an axiom of plural logic here, alongside comprehension (2.7) and indiscernibility (2.8).

Now (6) follows from (10) and the definition (9). Inversely, (10) follows from (6) and the definition (8). As for (7), the acceptance of indiscernibility (2.8) as a scheme means the acceptance of all its instances, whatever notations are added to the language. We have now added one new notation, the α-symbol, as we have the following new instance of (2.8) for it:

(11) $xx == yy \rightarrow \alpha xx = \alpha yy$.

From (11) and the definition (9) (together with comprehension), (7) follows without the need to assume anything like it as an additional axiom.

The main point is that with an α-symbol and the sole additional axiom (10) for it, the plural-logical axiom of choice (2) can be deduced. The *ys* required by (2) are given as follows:

(12) $u \propto yy \leftrightarrow \exists\exists xx(\Phi(xx)\ \&\ u = \alpha xx)$.

The verification that (2) does follow from (10) given the definition (12) is left to the reader.

9 SET-THEORETIC MODELS FOR PLURAL LOGIC

So much for the axiomatization of set theory (by heredity, extensionality, separation, and reflection), starting from a suitable plural logic (with comprehension, indiscernibility, and choice). We have seen what the plural-logical perspective of Boolos can contribute to set theory in the style of Bernays: it enables us to do without classes, and naturally suggests a choice of primitives that allows the reflection principle to be expressed in a particularly simple form that makes it arguably an expression of Cantor's principle of limitation of size.

Before closing, I should mention what, conversely, Bernays-style set theory can contribute to Boolos's plural logic. The point to be made is one familiar to specialists, though in the idiom of second-order logic and class-set theories, rather than of plural logic. It has also been treated briefly in the idiom of plural logic in Boolos (1985), but for the sake of completeness, and because Boolos's treatment is rather compressed, and stops short of explicitly endorsing the reflection principle, let me review the matter here.

But first, since the point to be made pertains to a *model theory* for plural logic, a word of caution may be in order. Throughout philosophical logic, much mischief is caused by a double usage of the word "semantics." It is used on the one hand for models, like those provided by Tarski for singular or first-order logic, or by Kripke for modal logic; and it is used on the other hand for a theory of meaning. Confusion between these two usages is manifested in the literature in two different, complementary ways. On the one hand, if a model theory has not yet been developed for a given logical notion, it may be alleged that the notion is "meaningless" because it lacks a "semantics." On the other hand, once a model theory *has* been developed for a given logical notion, it may be alleged that problematic "ontological commitments" are implicit in use of the notion – for instance, that ontological commitment to "unactualized possible worlds" is implicit in the use of modal notions – because something like them appears in its "semantics."

Both types of objections could be raised against plural logic. On the one hand, I have not yet presented a model theory for plural logic, and when I do it may not immediately be found satisfactory, so it might be claimed that the meaningfulness of plural quantification is in doubt. On the other hand, when I do present a model theory and an argument that it is satisfactory, if that argument is accepted, then since the model theory will involve an apparatus of sets, it might be claimed that this shows that an "ontological commitment" to sets is implicit in the use of the plural.

Against the first objection I maintain that even if no one ever did present a satisfactory model theory for plural logic, the plural was in systematic use in natural languages long before model theory for anything had been born or thought of, and such long-standing systematic usages are meaningful if anything is. Against the second objection I maintain – in addition to Boolos's point about the Russell paradox in (3.10) and (3.11) and their analogues for classes – that the transition from plural language to set-theoretic language in the work of Cantor and his followers involved an intellectual struggle more difficult than would have been called for if the task had been merely one of making explicit something already implicit in ordinary language.

So much by way of warning against confusing model theory with "semantics" in the sense of a theory of meaning. Now for the model theory itself. Familiarity will be assumed with the usual notion, due to Tarski, of a model for first-order logic, though it may be well to begin with a quick outline even of this. To illustrate the principles involved, it will be enough to consider formulas with just one non-logical symbol, a two-place predicate R.

Under Tarski's definition, a model $\mathbf{M}$ would consist of a non-empty set M, the *universe* of the model, and a set R^M of ordered pairs of elements of M, the *relation* of the model. In order to define what it is for a formula without free variables to be true in the model, one must define what it is for a formula with free variables to be true in the model relative to an assignment of elements of the universe to its free variables. The details of the inductive definition will not be recalled, except to mention that at the base step, Ruv is true for the assignment of μ and ν to u and v if and only if the ordered pair (μ, ν) belongs to the set R^M; while at the induction step for the universal quantifier, $\forall u \Psi(u)$ is true if and only if $\Psi(u)$ is true whatever element μ of M is assigned to u. In (one version of) the usual symbolism:

(1) $\mathbf{M}, \mu, \nu \models Ruv$ iff $(\mu, \nu) \in RM$
(2) $\mathbf{M} \models \forall u\Psi(u)$ iff $\mathbf{M}, \mu \models \Psi(u)$ for all $\mu \in M$.

There is a standard extension of this notion of model to *monadic second-order logic*, where we have in addition to the variables $u, v, \ldots$ a second style of variable $X, Y, \ldots$ and an additional two-place logical predicate ε, and therewith a new kind of atomic formula exemplified by $u \varepsilon X$. All that is changed in the definition is that truth is now relative to an assignment not only of an element μ of M to each variable u of the old kind, but also of a *subset* Ξ of M to each variable X of the new kind. At the base step, $u \varepsilon X$ is true for the assignment of μ to u and Ξ to X if and only if μ is an element of Ξ. At the induction step for the universal quantifier, $\forall X\Psi(X)$ is true if and only if $\Psi(X)$ is true whatever subset Ξ of M is assigned to X. In symbols:

(3) $\mathbf{M}, \mu, \Xi \models u \in X$ iff $\mu \in \Xi$
(4) $\mathbf{M} \models \forall X\Psi(X)$ iff $\mathbf{M}, \Xi \models \Psi(X)$ for all $\Xi \subseteq M$.

Turning now to plural logic, it has what I will call an *official* model theory in which truth is defined relative to an assignment of an element μ of M to each singular free variable u, and a subset Ξ of M to each plural free variable xx. At the base step, $u \propto xx$ is true for the assignment of μ and Ξ to u and xx, respectively, if and only if μ is an element of Ξ. At the induction step for the universal quantifier, $\forall xx\Psi(xx)$ is true if and only if $\Phi(xx)$ is true whatever subset Ξ of M is assigned to xx. In symbols:

(5) $\mathbf{M}, \mu, \Xi \models u \propto xx$ iff $\mu \in \Xi$
(6) $\mathbf{M} \models \forall xx\Psi(xx)$ iff $\mathbf{M}, \Xi \models \Psi(xx)$ for all $\Xi \subseteq M$.

Obviously, the effect is just the same as if each plural variable xx were replaced by a variable X, and each atomic formula $u \propto xx$ by an atomic formula $u \varepsilon X$, and then the standard model theory applied to the resulting second-order sentence. In this official model theory for plural logic *validity* is defined in the usual way as truth in all models. The result is that the claim that some plural formula is valid will be expressible in the language *singular* set theory, since plurals are not used in the metalanguage in (5) and (6); indeed, such a claim will be equivalent to the claim that a corresponding second-order formula is valid in the standard model theory for second-order logic, making all the many known results about that standard model theory applicable to plural logic.

Two questions immediately arise about how satisfactory this official model theory is, and how validity in the official sense is related to validity in the intuitive sense of being true in all interpretations. The first question is whether it would not be more natural to define truth relative to an assignment of *some elements*, the ξs, of M to each free plural variable xx, with induction clauses as follows:

(7) $\mathbf{M}, \mu, \xi\xi \models u \propto xx$ iff $\mu \propto \xi\xi$
(8) $\mathbf{M} \models \forall xx \Psi(xx)$ iff $\mathbf{M}, \xi\xi \models \Psi(xx)$ for all $\xi\xi \in\in M$.

Boolos's answer is that indeed it would more natural to use plurals in the metalanguage, but that the official singular definition with (5) and (6) is in fact equivalent, given the comprehension axiom of plural logic and the separation axiom of set theory, to the more natural plural definition with (7) and (8). For on the one hand, by separation, for any elements, the ξs, of M, there is a subset Ξ of M such that a given element μ of M is an element of Ξ if and only if it is among the ξs. And on the other hand, by comprehension, for any subset Ξ of M, there are some elements, the ξs, of M such that a given element μ of M is among the ξs if and only if it is an element of Ξ.

The second question is whether it would not be more natural to define validity in terms, not of models whose universes must be sets, but of interpretations in a more general sense, in which the objects over which the variables range might well be *all* objects, or all pure sets, or all pure and hereditarily well-founded sets, or the like. (The analogous question arises even in the case of *first-order* logic, and was raised long ago by Kreisel (1967).) To specify an interpretation in this more general sense, one need only specify, by means of some condition (perhaps involving parameters) which objects the variables are to be understood as ranging over, and by another condition (again perhaps involving parameters), which ordered pairs of such objects are to be understood as R-related. The official definition would be a special case of this general notion, so that validity in the natural sense would imply validity in the official sense, but there would be a question about the converse implication.

An answer extrapolated from Boolos's remarks would be that indeed again the proposed definition would be more natural than the official one, but that again the official definition is equivalent, this time by the reflection principle, to the more natural definition. To see this, suppose a sentence Γ is *not* valid in the general sense, and consider conditions $\Theta(x)$ and $\Psi(x, y)$ specifying an interpretation in which Γ is false. Here $\Theta(x)$ indicates what objects the variables are to be interpreted as ranging over, and $\Psi(x, y)$ which ordered pairs of such objects are to be interpreted as being R-related. Now go through Γ and replace each of the items on the left below by the corresponding item on the right:

$\forall u \ldots$	$\forall u(\Theta(u) \rightarrow \ldots)$
$\exists u \ldots$	$\exists u(\Theta(u) \,\&\, \ldots)$
$\forall xx \ldots$	$\forall xx(\forall u(u \propto xx \rightarrow \Theta(u)) \rightarrow \ldots)$
$\exists xx \ldots$	$\exists xx(\forall u(u \propto xx \rightarrow \Theta(u)) \,\&\, \ldots)$
Ruv	$\Psi(u, v)$

Calling the resulting formula Φ, the fact that $\sim\Gamma$ was true under the given interpretation implies that $\sim\Phi$ is true. Now apply reflection to obtain a set t such that $\sim\Phi^t$ is true. Let M be the set of all elements of t for which the condition Θ^t holds, and R^M the set of all ordered pairs of elements of M for which the condition Ψ^t holds, thus obtaining a model **M**. A little thought shows that the fact that $\sim\Phi^t$ is true implies that $\sim\Gamma$ is true in **M**, showing that Γ is not valid in the official sense either.

So much by way of argument that the official model theory is satisfactory. I have mentioned that this model theory makes all the many known results about the standard model theory for second-order logic applicable to plural logic. One of these results is the *incompleteness*, or rather, the *incompletablity*, of the logic. Validity does not correspond to deducibility using comprehension, indiscernibility, and choice – the latter being a sufficient, but not a necessary, condition for the former – nor would the addition of further axioms enable one to capture validity either, so long as the set of axioms is recursive. For the sentences deducible from a recursive set of axioms form a recursively enumerable set, and it is known that the second-order sentences valid in the standard model theory do not.

That is the bad news. There is also, however, some good news implicit in the discussion above of why the official model theory is satisfactory. The good news is that it will never become necessary to add any logical axioms to the three we already have, even if we become convinced that something further not deducible from them is intuitively valid, because the effect of adding a new logical axiom can always be obtained by adding a new set-theoretic axiom instead, namely, the axiom asserting that the candidate logical axiom is valid in the official model theory. (In the converse direction, in many *but not all* cases adding a new set-theoretic axiom is equivalent to adding a new logical axiom, namely, in all those cases where the candidate set-theoretic axiom is equivalent to the assertion of the validity of some second-order sentence in the standard model theory. This is known to include very many candidate set-theoretic axioms that have been considered in the literature, but not all large cardinal axioms.) In this sense, the logical axioms we have are all the logical axioms we will ever need.

In closing, let me reiterate that what we needed to establish that the official model theory is satisfactory, and therewith the somewhat reassuring result just stated, were precisely the set-existence axioms of BB, separation and reflection. The plural logic of Boolos and the reflection principle of Bernays, though introduced independently, the one decades after the other, turn out to be ideally suited for each other. The combination of the two is a marriage made in heaven – or at least, in Cantor's paradise.

7
Logicism: a new look

I NEO-NEO-LOGICISM

After a quick review of the original Fregean logicist program, I would like to describe two recent revivals of logicist ideas – Richard Heck's predicativist logicism, and the late Richard Jeffrey's logicistico-formalism – and briefly suggest how the two might be combined.

Frege in his *Begriffsschrift* (1879/1967) presented a deductive system of second-order logic – with the first-order entities called "objects" and the second-order "concepts" – including an absolutely unrestricted axiom of *Comprehension*, as follows:

(1) $\exists X \forall x (Xx \leftrightarrow \phi(x))$

together with its analogues for two-, three-, and many-place relational concepts or relations. The axiom of *Extensionality*, in the following form:

(2) $X \equiv Y \rightarrow (\phi(X) \leftrightarrow \phi(Y))$

is then provable using the following definition of coextensiveness, "the analogue of identity" for concepts:

(3) $X \equiv Y \leftrightarrow \forall z (Xz \leftrightarrow Yz)$.

Since (2) implies the uniqueness of the concept whose existence is asserted by (1), we may speak of *the* concept under which fall all and only those objects x for which $\phi(x)$ holds, or for short, the concept of being an x such that $\phi(x)$, for which I will write $\langle x\!: \phi(x)\rangle$.

Frege added, informally in the *Grundlagen* (1884/1950) and formally in the *Grundgesetze* (1893/1903) for the purposes of the derivation of arithmetic from logic, the infamous Basic Law V, the assumption that to each concept X is associated an object $^*X$, its *extension*, in such a way that the extensions of two concepts are identical if and only if the concepts are coextensive:

(4) $^*X = {}^*Y \leftrightarrow X \equiv Y$.

We may then introduce set-theoretic notation as follows:

(5) $\mathrm{Set}(y) \leftrightarrow \exists Y(y = {}^*Y)$
(6) $x \in y \leftrightarrow \exists Y(y = {}^*Y \,\&\, Yx)$
(7) $\{x: \phi(x)\} = {}^*\langle x: \phi(x)\rangle$.

Frege then defines *equinumerosity* in the familiar Cantorian manner:

(8) $X \approx Y \leftrightarrow \exists R(R$ is a bijection between X and $Y)$

where I have not bothered to write out the definition of bijection in purely logical terms. He then defines *number* thus:

(9) $\#X = \{y: \exists Y(y = {}^*Y \,\&\, X \approx Y)\}$.

Using this definition he derives what has come to be called (on account of a passing allusion on Frege's part to Hume) "Hume's Principle" or HP:

(10) $\#X = \#Y \leftrightarrow X \approx Y$.

He then proceeds to define zero, successor, and *natural* number, and to derive the Peano postulates.

Russell (1902) famously found a contradiction in Frege's system. It is almost too well known to bear repeating, but it needs to be pointed out that the contradiction about the set of all sets that are not elements of themselves uses *both* comprehension for a formula with a bound second-order variable *and* the assumption of the existence of extensions:

(11) $\{x: \mathrm{Set}(x) \,\&\, x \notin x\} = {}^*\langle x: \exists X(x = {}^*X \,\&\, {\sim}Xx)\rangle$.

To block the paradox, therefore, one might either restrict (1), or restrict or replace (4).

The first option was not seriously explored until recently. To be sure, Russell professed a "vicious circle principle" banning definitions of concepts by conditions involving quantification over concepts, or *impredicative* definitions as they came to be called. But he also introduced an "axiom of reducibility" whose effect was to circumvent such predicativity restrictions, so that as Ramsey (1925) observed he might as well not have imposed them in the first place.

The second option is what saves Russell from the contradiction. His "no classes theory" rejected (4) altogether, and with it Frege's approach to logicism, on which numbers are objects or first-order entities. (For Russell they are third-order entities, and he has to make essential use, as Frege did not, of higher-order logic; also, since unlike Frege he cannot prove that there are infinitely many objects, he has to assume so as the "axiom of infinity.") Contemporary neo-Fregeanism, whose continuous history begins (though there were significant precursors) somewhat over two decades ago with Crispin Wright's *Frege's Conception of Numbers with Objects* (1983), retains (1) and rejects (4), but in place of the latter assumes (10) as axiomatic, thus making numbers into objects after all.

The key technical result about this approach was given precise formulation and rigorous proof by the late George Boolos (1987), who showed that second-order logic with HP, or *Frege arithmetic*, and second-order Peano arithmetic, or second-order logic with the Peano postulates, are interpretable in each other and hence equiconsistent. The direction of the interpretability of second-order Peano arithmetic in Frege arithmetic he called *Frege's theorem*. There are serious questions how much more of mathematics beyond second-order arithmetic one can get on a natural extension of such an approach. (Kit Fine (2002) has a natural-seeming extension that gives third-order arithmetic.) There is also a question whether (10) is close enough to being a "logical" principle (as Frege thought (4) was) to justify neo-Fregeans in calling their position "neo-*logicism*."

The latter kind of question seems less of a issue with Richard Heck's *neo-neo-logicism*, which takes the opposite approach of retaining (4) but restricting (1), assuming only *predicative* comprehension. Heck (1996) explored what one would have been left with if one had modified Frege's system *only* by imposition of a "vicious circle" restriction, without any "reducibility axiom" to cancel it, and without the further "no classes" restriction. He was able to show (building on early work of Terence Parsons (1987)) that a system of this kind was consistent. He was also able to show that it is sufficiently strong to interpret Raphael Robinson's system Q of minimal arithmetic. Now while that system appears very weak, an idea of Robert Solovay, further pursued by Edward Nelson and then a number of others, has shown that ostensibly much stronger theories can be interpreted in Q. See Hájek and Pudlak (1998) for details. All these stronger theories can then be interpreted indirectly (*via* Q) in Heck's system.

Further refinements of these results are possible in several directions. First, interpretability of Q can be proved for weaker predicative systems than Heck's. In the very simplest such system, which I call PV (with P for "predicative" and V pronounced "five"), one has comprehension in its simplest predicative form:

(12) $\exists X \forall x(Xx \leftrightarrow \phi(x))$ *provided there are no bound concept variables in* ϕ

and in addition Law V in its original form (4). With the definitions of set-theoretic notions as before we get a little bit of set theory. We of course get the axiom of extensionality for sets

(13) $\mathrm{Set}(x)\ \&\ \mathrm{Set}(y)\ \&\ \forall z(z \in x \leftrightarrow z \subset y) \rightarrow x = y$

from the axiom of extensionality for concepts when sets are introduced as extensions of concepts. We also easily get empty sets, singletons, pairs and so on, thus:

(14) $\varnothing = \{x\colon x \neq x\}$
(15) $\{u\} = \{x\colon x = u\}$
(16) $\{u, v\} = \{x\colon x = u \vee x = v\}$.

Here the defining formulas have no concept variables at all. When we recall that we also may allow *free* concept variables as parameters, we see that we also get complements, intersections, and unions:

(17) $-^{*}X = \{x\colon \sim Xx\}$
(18) $^{*}X \cap {}^{*}Y = \{x\colon Xx\ \&\ Yx\}$
(19) $^{*}X \cup {}^{*}Y = \{x\colon Xx \vee Yx\}$.

One consequence of (14) and (18) is worth mentioning – the axiom of *adjunction*:

(20) $\forall x(\mathrm{Set}(x) \rightarrow \forall y \exists z \forall w(w \in z \leftrightarrow w \in x \vee w = y))$.

This guarantees the existence of $x \cup \{y\}$ for any set x and any object y. That (13)–(19) and their consequences such as (20) are in effect *all* we get can be proved (for the *cognoscenti*: by using elimination of quantifiers for monadic first-order logic with identity).

This does not look much like set theory, but in the famous little book *Undecidable Theories* by Tarski, Mostowski, and Robinson (1953), which first introduced Q, a joint result of Tarski and Wanda Smielew is mentioned without proof, to the effect that Q is interpretable in the set theory whose axioms are extensionality, empty set, and adjunction. In order to get an interpretation of Q we may take $0 = \varnothing$ and for successor use either the Zermelo definition $x' = \{x\}$ or the von Neumann definition $x' = x \cup \{x\}$. Interestingly enough, about ten years ago Franco Montagna and Antonella Mancini (1994) showed that Solovay–Nelson methods for interpreting other theories in Q can be adapted to prove the Szmielew–Tarski theorem about the interpretability of Q in another theory.

Second, in the case of stronger systems of arithmetic interpretable in Q and hence by the Smielew–Tarski theorem indirectly interpretable in PV, it is sometimes easier to prove interpretability in PV *directly*. For the most important specific example, a Δ_0-formula in the language of arithmetic is one containing only bounded quantifiers $\forall x < y$ and $\exists x < y$. A difficult argument of Alex Wilkie (as reported in Hájek and Pudlak, 1998) shows that the system called $I\Delta_0$, with the principle of mathematical induction (if $\phi(x)$ holds for zero and for the successor of any number for which it holds, then it holds for all numbers) for Δ_0-formulas, can be interpreted in Q. But A. P. Hazen showed that we can get interpretability of $I\Delta_0$ in PV *without* Wilkie's difficult argument. See Burgess and Hazen (1998) for details.

Third, one can actually go beyond $I\Delta_0$, which allows only for the operations of addition and multiplication, to a system which allows also exponentiation, if one works in a slightly stronger predicative Fregean theory, which I will call P^2V. In this theory there are objects x, degree-zero concepts X^0, and degree-one concepts X^1. A formula is of *degree* zero if it contains no degree-one variables, and no *bound* degree-zero variables, and is of *degree* one if it contains no bound degree-one variables. We then have two forms of comprehension, for the two degrees:

(21) $\exists X^0 \forall x(X^0 x \leftrightarrow \phi(x))$ *for* ϕ *of degree zero*
(22) $\exists X^1 \forall x(X^1 x \leftrightarrow \phi(x))$ *for* ϕ *of degree one.*

P^2V, with axioms (20), (21), and (4), can interpret the theory known as $I\Delta_0(\exp)$, which in a convenient equivalent of its usual formulation may be taken to have the following axioms:

(23) $0 \neq x'$
(24) $x \neq y \rightarrow x' \neq y'$
(25) $\sim x < 0$
(26) $x < y' \leftrightarrow x < y \vee x = y$
(27) $x + 0 = x$
(28) $x + y' = (x + y)'$
(29) $x \cdot 0 = 0$
(30) $x \cdot y' = (x \cdot y) + x$
(31) $x \wedge 0 = 0'$
(32) $x \wedge y' = (x \wedge y) \cdot x$
(33) $(\phi(0) \,\&\, \forall x(\phi(x)) \rightarrow \phi(x')) \rightarrow \forall x \phi(x)$
provided ϕ *is a* Δ_0*-formula.*

A formula like the conclusion of (33), consisting of universal quantifiers preceding a Δ_0-formula is called a Π_1-formula. Many important theorems of number theory (including the Fermat–Wiles theorem) have this form. It is known that quite a bit of mathematics can be developed in such a system as the above (and others known to be interpretable in or conservative over it). Harvey Friedman has conjectured that every Π_1-theorem published in the *Annals of Mathematics* can be proved in such a system.

The refinements of Heck's work that I have been discussing so far for the most part go back to a joint paper by Hazen and myself. That paper left open how much further one can go. It turns out that, unfortunately, one is not going to get in this way the whole of first-order arithmetic, or even all of so-called primitive recursive arithmetic, PRA, which is generally accepted, following William Tait, as a formalization of *finitist* mathematics. This is so even if one goes beyond P^2V to P^3V and so on, defined in the obvious way, or even $RPV = P^{\infty}V$, the union of all the P^nV, which amounts to so-called *ramified* predicative second-order logic plus Law V. This is on account of considerations related to Gödel's second incompleteness theorem, together with the fact that there is a *finitist* consistency proof for RPV. (The original consistency proofs of Parsons and Heck were model-theoretic rather than proof-theoretic in character, and infinitistic rather than finitistic.) This is the main new result in this area to be found in my little book *Fixing Frege* (2005b) where technical details and bibliographical references for all the material described so far can be found. (Important improvements have since been obtained in forthcoming work by Mihai Ganea and by Albert Visser.) To sum up so far, predicative logicism provides a foundation for a respectable modicum of arithmetic, but nothing anywhere near the whole of classical mathematics.

2 LITE LOGICISM

Lite "beer" is a fluid with approximately the taste of a mixture of 50 per cent real beer and 50 per cent soda water. Lite "logicism" is something brewed up by my late colleague Richard Jeffrey in his last years (1996, 2002), with the approximate composition 50 per cent logicism plus 50 per cent formalism. Though this recipe does not, perhaps, make it sound very appetizing, I myself on tasting it have found it considerably more palatable than I had expected, and I would like to say enough about it to tempt some of you to take a sip. The leading idea can be brought out by contrasting the following *Hilbert proportion*:

(1) computational : mathematics :: empirical : physics

which is the leading idea behind formalism, with the following *Jeffrey proportion*:

(2) logical : mathematics :: empirical : physics.

So let me begin with Hilbert in interpreting whom I follow (Weyl, 1944).

With both Hilbert and Jeffrey we have a view of mathematics self-consciously modeled on a certain philosophy of physics (more popular perhaps in Hilbert's day than in Jeffrey's). On this view the theoretical portions of physics are only there to imply empirical laws: universal generalizations whose instances are empirically decidable. Now it is an immediate consequence of the definition of empirical decidability that theoretical physics and empirical laws cannot imply any empirically decidable sentences that could not be discovered directly by observation. But they can yield such sentences more quickly, as *predictions* of experiences future rather than records of experiences past. A better description of the philosophy of physics in question is that it takes physics to be nothing but a giant engine for generating empirical predictions. While the "nothing but" is controversial, it is comparatively uncontroversial that the *data* to which physics is responsible are empirical, and that *empirical fruitfulness* must be demanded. It is demanded of higher and higher theories that they should continue to yield more and more empirical predictions.

On Hilbert's view the theoretical (or as he called them "ideal") portions of mathematics are only there to imply universal laws whose instances are computationally decidable sentences (which he considered the only "real" sentences). In modern terminology these would be Π_1-sentences. Now again it is an immediate consequence of the definition of computational decidability that theoretical mathematics and Π_1-sentences cannot imply any computationally decidable sentences that could not be discovered directly by calculation. But again they can yield such sentences as "predictions" (for instance, the commutative law of multiplication predicts that the results of multiplying two hundred-digit numbers in one order and in the opposite order will be the same far more quickly than we can verify as much by tedious calculations). Hilbert thought that higher and higher mathematical theory would be *computationally fruitful*, yielding new computational predictions, in the sense that it would yield Π_1-sentences more quickly, but not in the sense that it would outright yield more Π_1-sentences. On the contrary, his program was to try to convince the finitist of the *reliability* of classical mathematics, through proving by finitist means that for any Π_1-sentences having a classical proof, it is possible in principle, though perhaps not feasible in practice, to produce a finitist proof.

Gödel's incompleteness theorems showed Hilbert was wrong: classical mathematics cannot be finistically proved to be reliable or even consistent; we must be content with inductive evidence on these points. This is the negative side of the coin whose positive side is that higher and higher theories do not just yield quicker proofs of Π_1-sentences that could be proved, albeit perhaps very much more slowly, in lower theories, but yield outright new Π_1-sentences, so that we have computational fruitfulness in a stronger sense than Hilbert expected. All this is because a higher theory can prove the consistency of a lower theory, as the lower theory itself cannot, and because consistency can be expressed as a Π_1-sentence.

Though I have so far spoken only of Π_1-sentences, there are important distinctions among them. Hilbert accepted Π_1-sentences involving arbitrary primitive recursive functions, including every function in the sequence addition, iterated addition or multiplication, iterated multiplication or exponentiation, iterated exponentiation, and so on. From certain philosophical points of view, however, one might wish to stop the series with exponentiation, or even with multiplication. (Recall, in particular, that the simplest form of predicative logicism gave only addition and multiplication, while the ramified form gave exponentiation but not much more.) In that case, the question of computational fruitfulness, of whether higher and higher theory does more and more computational work, would have to be reopened. Do we actually get new Π_1-sentences *involving only addition and multiplication and exponentiation*? Do we get new Π_1-sentences *involving only addition and multiplication*? (We do not get any involving only addition; for the *cognoscenti*, this is a consequence of Pressburger's theorem.) These questions are non-trivial, but answers are known. An affirmative answer to the former question is implied by the work of Julia Robinson (building on the work of Martin Davis and Hilary Putnam), and an affirmative answer to the latter question is implied by the work of Yuri Matiyasevich that builds thereupon, for an exposition of which see Matiyasevich (1993).

No logicist of any stripe can be satisfied with an approach that, like Hilbert's as described so far, takes computation to be an end in itself, without regard to applications. Even Hilbert recognized that some connection must be made between the use in pure mathematics of numerals as *nouns*, denoting numerical objects on which we can perform arithmetical operations, and the doubtless historically far older use in applications of numerals as *adjectives*, in numerically definite quantifications. The computational fact that $2+2=4$ is obviously somehow connected with the logical fact that if two sheep jumped the fence in the morning and two sheep jumped the fence in the evening, and no sheep jumped both morning and evening, then four sheep altogether jumped the fence morning or evening. But Hilbert gave no *formal* account of the connection.

Why do I call the fact about the sheep a *logical* fact? If we think of first-order logic (with identity) as enriched by (definable) numerical quantifiers of the type

(3.1) $\exists_1 xAx \leftrightarrow \exists xAx$
(3.2) $\exists_2 xAx \leftrightarrow \forall y\exists x(x \neq y \mathbin{\&} Ax)$
(3.3) $\exists_3 xAx \leftrightarrow \forall z\forall y\exists x(x \neq z \mathbin{\&} x \neq y \mathbin{\&} Ax)$

and so on, along with the quantifiers

(4) $\exists_m!xAx \leftrightarrow \exists_m xAx \mathbin{\&} \sim\exists_{m+1} xAx$

then the fact about the sheep becomes an instance of the general first-order logical law

(5) $\exists_2!xAx \mathbin{\&} \exists_2!xBx \mathbin{\&} \sim\exists x(Ax \mathbin{\&} Bx) \rightarrow \exists_4!x(Ax \vee Bx)$.

Any variety of logicist or set-theoretic approach will supply what Hilbert does not, a *systematic* way of connecting arithmetical facts with logical laws, through the characterization of arithmetical relations and operations that Fregean and Russellian logicism share with (or rather, borrow from) the Cantorian theory of cardinal numbers. To recall those characterizations, if *a* and *b* and *c* are respectively the number of *A*s and *B*s and *C*s, then we have the following characterizations of arithmetic notions by existential second-order formulas, wherein *R* is two-place and *S* is three-place:

(6.1) $a \leq b \leftrightarrow$
$\exists R$(*R* gives a bijection between the *A*s and some of the *B*s)

(6.2) $a + b = c \leftrightarrow$
$\exists R$(*R* gives a bijection between the *A*s and some of the *C*s and *R* gives bijection between the *B*s and the rest of the *C*s)

(6.3) $a \cdot b = c \leftrightarrow$
$\exists S$(*S* gives a bijection between the pairs consisting of an *A* and a *B* and the *C*s)

where "bijection" is to be written out in logical terms in the usual way. If the numbers involved are finite we also have

(7.1) $a < b$
$\exists R$(*R* gives a bijection between the *A*s and some but not all of the *B*s)

(7.2) $a \wedge b = c \leftrightarrow$
$\exists S(\forall x(Cx \rightarrow S(-, -, x)$ gives a function Sx from the Bs to the As)
& $\forall x \forall w(Cx$ & Cw & $x \neq w \rightarrow Sx$ and Sw are distinct)
& $\forall x \forall y \forall z(Cx$ & By & $Az \rightarrow \exists w(Cw$ & $Sw(y) = z$ and otherwise Sw agrees with Sx)

To illustrate by example how these characterizations provide a bridge between the computational and the logical, consider the simple law

(8) $\forall a \forall b \sim(a < c$ & $c \leq b$ & $b \leq a)$

and any particular numerical instance, say that for $c = 17$:

(9) $\forall a \forall b \sim(a < 17$ & $17 \leq b$ & $b \leq a)$.

Using the definitions of the numerical quantifiers and characterizations of arithmetical relations and operations above, this yields

(10) $\forall A \forall B \sim(\sim\exists_{17} x Ax$ & $\exists_{17} x Bx$ &
$\exists R(R$ gives a bijection between the Bs and some of the As))

which is equivalent to

(11) $\forall A \forall B \forall R \sim(\sim\exists_{17} x Ax$ & $\exists_{17} x Bx$ &
$\exists R(R$ gives a bijection between the Bs and some of the As))

which amounts to an assertion of the validity of the first-order scheme obtained by dropping the initial universal quantifiers in (11), or equivalently and more naturally to the validity of the first-order scheme

(12) $\sim\exists_{17} x Ax$ & $\exists_{17} x Bx \rightarrow$
$\sim(R$ gives a bijection between the Bs and some of the As).

One instance of (12) would be: if there are fewer than 17 pigeonholes and at least 17 pigeons, then it cannot be that each pigeon goes into one and only one hole, with only one pigeon per hole. The general law (8), by taking more and more specific numerical instances, yields more and more pigeonhole principles.

Hilbert's idea was that just as physics is "empirical" *not* in the sense that all its theoretical concepts admit definitions in terms of observation, but in the sense that its *data* are empirical, so is mathematics "computational" *not* in the sense that all its theoretical concepts admit definitions in terms of calculation, but in the sense that its *data* are computational. Jeffrey's idea is to substitute "logical" for "computational" here. Note that by the Gödel completeness theorem, no first-order logical law can be derived from arithmetic or via arithmetic from higher mathematics, in the manner sketched above, that cannot in principle be derived by textbook methods, though perhaps only very much more slowly. It is a question of getting results more quickly, as logical "predictions."

Now even in the case of pigeonhole principles, which Sam Buss has shown can be proved relatively quickly by textbook methods, proof by instantiation of (8) will be quicker for large values of *c*. But in this particular example, the Π_1-sentence (8) involved is one provable already in $I\Delta_0$. A question of *logical fruitfulness* now arises, which Jeffrey did not find time to consider, the question whether higher and higher theory in mathematics does any logical work, by yielding more and more logical "predictions," by yielding more and more Π_1-sentences *that in turn yield logical laws*. It turns out that the answer is affirmative, as a slight extension or corollary of the work of Robinson and Matiyasevich alluded to earlier. This is the one formal result in this area to be found in my paper "Protocol sentences for lite logicism" (Burgess, forthcoming), where technical details and bibliographical references for all the material described so far can be found.

A combination of the two new forms of logicism I have been discussing can now be contemplated. On such a combined view, the "real" mathematics, including the characterizations needed to derive logical laws from arithmetic results, and the most basic arithmetical results themselves, would be provided by predicative logicism, perhaps RPV or alternatively PV. The attitude towards "ideal" mathematics would be that of lite logicism: it is justified by, on the one hand, inductive evidence that its logical predictions are reliable, and on the other hand, by a metatheorem to the effect that higher and higher theory is fruitful in the sense of yielding more and more logical predictions.

I do not on the present occasion wish to *advocate* this form of logicism in any stronger sense than commending it to the reader's attention as worthy of consideration. For the combination to be completely satisfactory, we would need to know that the metatheorem in question can actually be proved in "real" mathematics, RPV or PV. For RPV this is almost certain, though I cannot claim to have dotted every *i* and crossed every *t*. Even leaving this question unresolved, among various other loose ends, it will I hope be clear that the potential of logicism is far from having been exhausted even today, well over a century after it was first introduced by Frege.

PART II

Models, modality, and more

8

Tarski's tort

I DEFINABILITY IN DISREPUTE

While what Alfred Tarski labeled his "semantic conception of truth" has been much discussed, one topic that has *not* received all the attention it deserves is his choice of that label. It is this comparatively neglected aspect of Tarski's conception that I wish to address here. But first a word about the situation prior to Tarski.

I begin with a result I learned as a fifteen-year-old student in a summer mathematics program for high-school students run by the late Arnold Ross: the theorem that every natural number is interesting. The proof is by contradiction. Suppose that not every natural number is interesting. Then the set of uninteresting natural numbers is non-empty. So by the well-ordering property of the natural numbers, it must have a smallest element *n*. But if *n* is the smallest uninteresting natural number, then *n* is interesting for that very reason. Thus we have a contradiction, establishing that our original hypothesis was false, and that every natural number is interesting after all. But, of course, some numbers do appear completely uninteresting to most of us. I suppose a so-called dialethist might claim that here we have yet another example of a true contradiction, but the more usual reaction to this bit of adolescent mathematical humor is that "interesting" is too vague or ambiguous, too subjective or relative, a concept to be admissible in mathematical reasoning. And when Alfred Tarski was beginning his mathematical career, most mathematicians held essentially the same opinion about the concept of truth. In order to understand what Tarski was up to with his truth definition, one needs to keep ever in mind this historical fact.

The first suspect notion that engaged Tarski's attention was not that of truth, but rather that of definability. That notion belongs to the same family as the notion of truth, since an object is *definable* if there is a condition true of or satisfied by it and it alone. And eighty years or so ago this notion of definability was in as much disrepute as the notion of truth itself. Thus Tarski begins his first relevant paper (Tarski 1931), on the notion of definability for sets of real numbers, by remarking "Mathematicians, in general, do not like to operate with the notion of definability; their attitude towards this notion is one of distrust and reserve." And he goes on to concede "The reasons for this aversion are quite clear and understandable." Indeed, the grounds for mathematicians' suspicions about definability are not far to seek. The emergence of Russell's and other set-theoretic paradoxes had put mathematicians in mind of some of the paradoxes propounded by ancient philosopher–logicians – Poincaré, as I recall, somewhere explicitly mentions Zeno the Eleatic and the school of Megara in this connection – beginning with the paradox of the liar. And soon people began inventing modern paradoxes of a similar stripe. The closest in spirit to Russell's paradox of the set of sets not elements of themselves was Grelling's heterological paradox, about adjectives or adjectival phrases not true of themselves. Paradoxes of this type had an important influence, in that they helped convince Russell and others that what was responsible for the set-theoretic paradoxes was not the assumption that actual infinities exist – the heterological paradox has nothing to do with infinity – but a kind of self-reference or vicious circularity that came to be called impredicativity.

There were also paradoxes about definability: Berry's, Richard's, and above all König's. This last was especially important because while no one had ever taken Berry's or Richard's arguments to be anything but ingenious sophisms, König and presumably at least two others (the referee and the editor who accepted his note (König 1905/1967) for publication), took his argument to be a legitimate proof, or more precisely, a legitimate disproof of the hypothesis that the continuum can be well ordered. The argument, it will be recalled, is that supposing there exists a well-ordering *W* of the continuum, we may consider the set of real numbers that are definable in terms of *W* (including those that, like real number zero, for instance, are definable even *without* mentioning *W*). Since by one theorem of Cantor there are only countably many finite strings of letters of the alphabet to serve as definitions, this set must be countable. Since by another theorem of Cantor there are uncountably many real numbers, the complement of this set must be non-empty. But then, since *W* is by hypothesis a well-ordering, it must have a *W*-least element: the *W*-least real number not definable in terms of *W*. But this last description provides a definition of the number in question (in terms of *W*), thus yielding a contradiction, and refuting the supposed theorem of Zermelo.

It was, however, not just on account of this paradox of König's that the repudiation of definability as an unmathematical notion was especially firm on the part of the defenders of set theory. For there were other, more direct arguments than König's against Zermelo's axiom of choice, based on the assumption that the existence of a set or relation or function depends on its being definable. In replying to these arguments, the defenders of set theory emphasized the lack of mathematical precision in the notion of definability. Thus Hadamard, in the exchange with other French analysts on the axiom of choice (Baire *et al.* 1905), makes this point and goes so far as to say the notion of definability belongs to psychology rather than mathematics.

And yet Tarski saw that the notion of definability had important *mathematical* applications, as becomes especially clear in the various follow-ups to Tarski (1931/1983b) from which emerged the so-called Tarski–Kuratowski algorithm, which allows one to compute the topological complexity of point-sets in the line or plane or space (open, closed, F_σ, G_δ, Borel, analytic, co-analytic, projective) by consideration of the logical complexity (number of alternation of quantifiers when reduced to prenex form) of the condition that defines the set. The ideas in this short paper, though of the sort that once absorbed come to seem obvious, were of the greatest importance, especially after Addison clarified the connection between the hierarchies of interest to topologists and the hierarchies introduced, also in the first half of the 1930s, by the recursion theorist Kleene. These ideas opened the door to the application of ever more sophisticated logical techniques to the descriptive theory of point-sets, eventually leading to the explanation by Gödel, Cohen, and their successors of why so many problems in descriptive set theory had resisted solution – they are undecidable on the basis of the conventional axioms of set theory – and in the work of Woodin and his predecessors, showing how a satisfactory solution was obtainable on the basis of large cardinal axioms.

It may be noted that Tarski in these early papers was concerned with something more general than the definability of an element of a domain (which was what was at issue with the Berry, Richard, and König paradoxes), namely, the definability of a subset of the domain, which is a matter of there being a condition satisfied by all and only the elements of that subset (definability of an element reducing to a special case, the definability of its unit set). If one-, two-, three-, and many-dimensional sets are in question, it is necessary to consider satisfaction for conditions with one-, two-, three-, or many variables. And as Tarski notes, the notion of truth is simply the degenerate or dimension-zero case of the notion of satisfaction. Thus there is a direct line between the earlier papers, which defined satisfaction for a single, specific interpreted language, and the great *Wahrheitsbegriff* paper, which defined it for all interpreted languages of a given kind. What concerned Tarski in that later paper is the same thing that had concerned him in his earlier papers, namely, the rehabilitation of a notion in disrepute among contemporary mathematicians.

2 MATHEMATICAL DEFINITION VS LINGUISTIC ANALYSIS

Tarski in his great paper on truth (Tarski 1935/1983c) was not interested in determining the meaning of the word "true." He thought he already had a *partial* understanding sufficient to determine what the *extension* of "true" was supposed to be. This understanding is expressed in Convention T, or rather, in his laying down Convention T as a criterion of "material adequacy." And he repeatedly tells us that he has no interest in going further and determining the *intension* of "true." The other requirement he states for a truth definition, that of "formal correctness," has nothing to do with fidelity to the intuitive sense of the term, but merely means mathematical rigor, which is of course essential if the suspicions prevalent among mathematicians about the notions with which he is concerned are to be allayed. Material adequacy and formal correctness are his only official requirements, though naturally he is also interested in the usefulness of notions he is defining, beyond the applications he has already made of them in his earlier papers on definability. To repeat, there is no requirement of fidelity to the intuitive, pre-theoretic sense of "true," beyond conformity to Convention T.

Of course, mathematicians in propounding definitions of mathematical terms for words already in extra-mathematical use are generally even less interested than Tarski was in their ordinary meanings. For they care no more about extensional than about intensional agreement between the technical sense being introduced and the ordinary sense, and they do not lay down any criteria of "material adequacy" based on their (total or partial) understanding of the word in its extra-mathematical sense. Thus mathematicians feel no compunction whatsoever in speaking about Hilbert "space," though it certainly is not the kind of "space" that, say, astronauts travel around in. But there is also a less superficial sense in which mathematicians are unconcerned with meanings. For they are not really concerned even with the meaning of the term in its *technical* sense, at least not if "meaning" is understood in the same fine-grained way it is understood by lexicographers propounding definitions or philosophers propounding analyses. This deeper kind of indifference to meaning, or at least interest only in a very coarse-grained kind of "meaning," is best illustrated by an example.

An analysis text may define the constant *e* using the well-known representation as the limit of a sequence:

(1) $e = \lim_{n\to\infty} (1 + n^{-1})^n$.

Thereafter, all theorems about *e* in the text will refer back to this definition, or to previous lemmas based on this definition. This includes, for instance, the well-known representations as the sum of a series:

(2) $e = 1/0! + 1/1! + 1/2! + 1/3! + 1/4! + \ldots$

Surely it is in this practice of fixing a definition and referring all later results back to it that we should look for the origin of Frege's notion that in a properly constructed scientific language, each name should be associated with a single, fixed definition.

The language of the mathematical community, however, does not conform to this requirement. For while one textbook writer may take (1) as the definition and (2) as a theorem, another writer may do the reverse. Even a mathematician taught originally from a textbook using one of these approaches may him- or herself, when he or she comes in due course to write a textbook, adopt the other. Moreover, even if a sizable majority favors one approach, which is as may be, they will admit the legitimacy of the other. There is no question of insisting on one definition as the true, original one. It would be wrong to say that mathematicians do not care about distinctions between definitions that are mathematically equivalent, or even to say that they do not care about distinctions between those that are *provably* equivalent, or even to say that they do not care about distinctions between those that have actually been *proved* to be equivalent, if the proof is long and difficult and involves advanced ideas. But they *are* indifferent to distinctions between definitions for which there are comparatively well-known, short, simple, elementary proofs of equivalence. (1) and (2) are certainly not synonymous in any reasonable sense of "synonymous," but they are equally acceptable as "definitions" of *e*. It is in this situation that we find the origin of the notion expressed in Quine (1936) and elsewhere that definitional status is local and transitory – a notion at the root of his ultimate rejection of the analytic/synthetic distinction.

This same kind of indifference to meaning may be found also in Tarski's paper, in addition to the kind of indifference to meaning I mentioned earlier. For as Wilfrid Hodges reminds us,[1] while Tarski's general definition of truth is applicable to arbitrary interpreted languages of a certain kind, for certain special interpreted languages another kind of definition of truth, very different in appearance from the general definition, is available: a definition of truth by elimination of quantifiers. The alternative definition is equally mathematically rigorous or "formally correct," and equally "materially adequate," so that it agrees in extension with the general definition applied to this specific case, though it is by no means synonymous with the general definition in any reasonable sense of "synonymous." Where such an alternative definition is available, Tarski is perfectly happy with it. Thus Tarski is unconcerned with the meaning of "true" in a double sense: first, he is unconcerned with pinning down precisely the meaning of "true" in ordinary language, or in insuring that "true" as a technical term will agree more than extensionally with "true" as an ordinary term; second, he is unconcerned with differences between alternative technical definitions, if these have been proved extensionally equivalent.

And if Tarski is doubly unconcerned with pinning down the meaning of "true," he is even less concerned with analyzing the meaning of any other word or symbol. A clause such as either of the following:

(3) (A and B) is true iff A is true and B is true
(4) $\mathrm{True}(A \wedge B) \leftrightarrow \mathrm{True}(A) \wedge \mathrm{True}(B)$

emphatically cannot be construed as telling us the meaning of the word "and" or the symbol "∧." For in Tarski's original set-up, the "object language" for which truth is being defined is *contained* in the metalanguage in which the proof is being given. That is why we see ay-en-dees on both sides of (3) and carets on both sides of (4). In order to understand the definition, one must understand the metalanguage, and that includes understanding the object language which is part of it, and therewith each of the words or symbols of the object language.

3 TARSKI'S TRIUMPH AND TRESPASS

It was not linguistic understanding but mathematical fruitfulness that Tarski sought with his definition, and in this he was very successful. For Tarski is the creator of the incredibly rich branch of mathematical logic

[1] In his "Tarski's truth definitions," in *Stanford Encyclopedia of Philosophy*, http://plato.stanford.edu/entries/tarski-truth/ (available only on-line).

known as the theory of models. Hodges, however, has noted how very slow Tarski was to advance beyond the 1933 definition of truth for sentences of an interpreted language to the definition of the notion that is central to the theory of models, namely, the notion of truth *in a structure* for sentences of an *un*interpreted language.

Formally the step is a very short one. An interpreted language is naturally thought of as an ordered pair consisting of an uninterpreted language and an interpretation. And an interpretation is simply a set, the domain, and an assignment of a relation or operation of the right number of places on it to each non-logical primitive. But that is essentially what a mathematical structure is: a set, the domain, and certain distinguished relations and/or operations on it, distinguished from each other by certain symbols associated with them. For instance, a *ring* is, first of all, a set with two binary operations, one written additively and one written multiplicatively. And formally, the step from a two-place relation between a sentence and an ordered pair consisting of an uninterpreted language plus an interpretation or structure to a three-place relation among a sentence, an uninterpreted language, and a structure or interpretation is a very short one. But though Tarski was in effect operating with the latter notion within a few years, he did not give an explicit definition until over two decades later. Indeed, it is oversimplifying to say that *he* gave the definition, since it appears in a joint paper with a student (Tarski and Vaught 1956).

This notion of a sentence of a given uninterpreted language being *true* in a structure, or conversely, of a structure being a *model* of a sentence of an uninterpreted language, has proved immensely useful both in applications to core mathematics (mainly abstract algebra) and in applications to the metamathematics of set theory. The notion is also needed to give a fully rigorous statement of the Löwenheim–Skolem and Gödel completeness theorems, which were proved before they were stated, so to speak.

Closely related to this last application was another application envisioned by Tarski, that of giving a definition of *logical truth*. Here, however, his work (Tarski 1936/1983d) is of more ambiguous status. Tarski himself pointed to one limitation of his definition of logical truth or truth by virtue of logical form alone as truth in all models: the definition presupposes a division of symbols and notions into logical and non-logical, of which Tarski was not to give an account until a late lecture, published only posthumously (though since much discussed). Kreisel (1967) pointed to another limitation: logical truth in an intuitive sense is truth in all interpretations in an intuitive sense, and is not restricted to truth in interpretations in the technical sense considered by Tarski, where the quantifiers must range over a *set* and not a proper class. For first-order logic, as Kreisel notes, the completeness theorem shows that it is enough to consider interpretations where the domain is a set. But for second-order logic (if one grants that it is *logic*) the assumption that holding in all interpretations and holding in all *set*-interpretations – the intuitive notion and the Tarski notion – coincide is not provable on the basis of the usual ZFC axioms for set theory, but is in effect a large cardinal assumption. But despite the limitations of Tarski's approach on this one issue, there is no denying that Tarski's definition of truth was immensely successful.

Yet Tarski's achievement was marred by a misdeed on his part that opened the door to considerable mischief. His violation did not rise to the level of felony or even misdemeanor – indeed, as my title suggests, it was a civil rather than a criminal wrong. It was a case of trademark infringement: his appropriating to his own use the linguists' term "semantics." Of course, it was not really an actionable offense at all, since academic disciplines, unlike business corporations, are not legal "persons" with standing to sue anyone, though in a way that is a shame. (It would be a very good thing if, for instance, the International Seismological Union could collect hefty punitive damages from any writer who uses "epicenter" as a fancy synonym for "center of activity.") And if Tarski's act were actionable, there would be, as in virtually all cases of this particular offense, something to be said for the defense. There was a minority usage of "semantics" for something other than a theory of the meanings of words. Indeed, there were several minority usages, and in "The semantic conception of truth" (Tarski 1944), he is quick to disassociate himself from one of them, the "General Semantics" of another Polish thinker, Count Alfred Korzybski.

But there can hardly be any question that what "semantics" conveyed and conveys to the mind of the general reader is a theory of *meaning*, which Tarski's theory most emphatically was not. By calling his theory "semantics," Tarski opened the door to endless misunderstandings on this point. There has been significant damage to logic arising from such misunderstandings, from confusion of model theory or "semantics" improperly so-called with meaning theory or "semantics" properly so-called. Needless to say, if one is careful, one can avoid the confusion even while keeping the double use of "semantics" by distinguishing *formal* from *linguistic*. But in general usage "formal semantics" is a case of oxymoron and "linguistic semantics" a case of pleonasm, and it would have been better not to create a situation where there is a call for distinguishing adjectives in the first place.

4 MODAL MUDDLES: SPURIOUS COMMITMENTS

Tarski's usage did create such a situation, or at least the decision by most of his successors to follow him in his usage of "semantics" has done so. For his usage, though not in fact followed by workers in first-order model theory, has generally been followed by philosophers, and even more so by computer scientists working in the field of modal, temporal, and related logics. Hence, for instance, the title "Semantical considerations on modal logic" for the paper Kripke (1963) in which the inventor of Kripke models finally belatedly published his model theory for modal logic. Not everyone who follows Tarski's usage is confused, of course, but everyone who does so encourages confusion, and there has been confusion enough, and this of two kinds: *spurious attributions of ontological commitment to commonplace locutions* and *unwarranted complacency about the intelligibility of dubious notions*. Let me take up the first of these phenomena first, illustrating by the case of tense logic.

In tense logic we have future-tense and past-tense operators F and P whose intended meaning is something like "it (sometime) will be the case that . . ." and "it (once) was the case that . . ." These behave syntactically like negation: they are one-place connectives. But it is not hard to see that we cannot treat them model-theoretically like negation. These parallel clauses:

(5) $\mathfrak{M} \models \sim$A iff it is not the case that $\mathfrak{M} \models$ A
(6) $\mathfrak{M} \models$ FA iff it will be the case that $\mathfrak{M} \models$ A

will *not* do because we are looking at *mathematical* models here, and mathematical facts about mathematical objects do not change. If a model does not satisfy a sentence now, there is no use waiting, for it never will.

The mathematical logician will naturally want to deal with this problem by adopting the kind of treatment of time that is used in mathematical physics, where time is in effect represented as an extra spatial dimension. This amounts to purging the language of tense, replacing sentences of the type, "It is raining" in the present tense by formulas of the type "It be raining at time *t*," where the "be" is tenseless, or rather by the instantiation of such a formula where a constant *c* for the present is put in for the variable *t*. Then instead of "It will rain," or in more stilted and stylized form, "It will be the case that it is raining," we get "There exists a time *t* such that *t* is future relative to *c* and it be raining at time *t*." This is, in Quinean terms, regimentation in first-order logic. Or if the logician does not insist on translating tense-logical sentences into first-order sentences quantifying over times, and introducing a relative futurity relation, the logician will anyhow, in presenting a *model theory* for tense logic, use exactly the notion of model one *would* get if one *did* translate into first-order terms and use Tarski-style model theory for the latter.

In other words, in place of a classical model, which at the sentential level is just an assignment of truth values to sentence letters, one has something more complicated: a domain X of times and a relative-futurity relation R on it, and an assignment to each a in X of a model of the classical kind, which is to say, an assignment of truth values to sentence letters. One then has to define not truth of a sentence in the model $\mathfrak{M}$, but satisfaction of a formula in the model $\mathfrak{M}$ by an element a of X. The relevant clauses then become as follows:

(5*) $\mathfrak{M} \models \sim A[a]$ iff it is not the case that $\mathfrak{M} \models A[a]$
(6*) $\mathfrak{M} \models FA[a]$ iff it there is a b such that aRb and $\mathfrak{M} \models A[b]$.

Now philosophically, thinking tenselessly raises a whole raft of new questions, and above all this one: according to our ordinary tensed ways of thinking, I always have been (as long as I have been around) and am always going to be (for as long as I am still around) three-dimensional. On this new tensed way of speaking, it seems that I be at different times, or if one prefers, in different temporary states of the world, or if one really wants to encourage confusion, in different temporary worlds. But how can I be in more than one (temporal) "place"? Should I conceive of myself as being four-dimensional, with various three-dimensional stages, "temporal segments" of me, at various times? Should I conceive of myself as being only at the present time, and merely having past and future counterparts? Quite a few puzzles arise. Nonetheless, it may be insisted that the tenseless way of thinking is more scientific, and perhaps even demanded by certain advances in physics during the last century.

That may be so, but even if it is, the old-fashioned tensed way of thinking is going to be around for a long time before the new-fangled tenseless way is universally adopted, and so we would do well to understand the relationship between the two. This is what tense logic, as developed by Prior (1967b), in effect does. Physical assumptions about the structure of time correspond naturally to assumptions about the structure or "frame" of times with the earlier–later relation (which is what relative-futurity amounts to), or in other words, different classes of models. Completeness theorems relate various axiomatic systems of tense logic to various classes of models, so that, for instance, the assumption that the earlier–later relation on times is *dense* corresponds to the assumption that $\mathsf{F}p \rightarrow \mathsf{FF}p$ is valid.

So far, so good. However, if one thinks of the model theory as a *meaning* theory for tense logic, one will be led to the idea that the tenseless way of thinking is not some new-fangled techno-scientific development of the last century, but rather has been what, despite superficial appearance to the contrary, our ordinary tensed ways of speaking "deep down" have meant all along. The notion of a durationless instant of time is not, on this view, a sophisticated, advanced, historically late posit, but rather is something that has been implicitly assumed since time immemorial. In a word – or rather, in two words – even our ordinary, old-fashioned, commonsense tensed ways of speaking and thinking are *ontologically committed* to instants of time – or worse, to "temporary worlds": "They aren't there on the surface, but they're there *in the semantics*." But in fact, their presence in the model theory – or if you must, in the *formal* "semantics" – is merely an artifact, an inevitable consequence of the fact that it is *mathematical*, and therefore *unchanging* models that we are considering. The "ontological commitment" here is entirely spurious, based on a simple fallacy of equivocation between misnamed formal "semantics" and properly linguistic semantics:

> *It's there in the models.*
> *So it's there in the semantics.*
> *So it's there in the meaning.*

Mathematical facts are necessary as well as permanent, and exactly parallel considerations apply to the modal case. Mathematical modelers will reach for a treatment of modality resembling that used in mathematical probability and statistics, with a space of "possibilities," or if you prefer "possible states of the world," or if you really want to encourage confusion "possible worlds." But this gives absolutely no reason whatsoever to suppose that when I say "I could have got stranded in an airport owing to a snow storm" I *mean* there is a possibility or possible state of the world or possible world where I, or a state or modal segment of me, or a counterpart of me, is stranded. The fact that what some people insist on calling "possible worlds" are there "in the semantics" tells us nothing about "ontological commitment" to them, if the "semantics" is only *formal* and not *linguistic*. Similar remarks apply to the supposed ontological commitment of plural logic to *sets* or *classes*, and there are other examples as well. The clash between the usage of "semantics" by Tarski and his followers and the majority use in the educated public is forever misleadingly suggesting existential implications that are not there.

5 MODAL MUDDLES: UNWARRANTED COMPLACENCY

It is a significant historical fact that the model theory for modal logics was worked out in the late fifties and early sixties, while the distinctions among logical demonstrability, logical validity, analyticity, *aprioricity*, and necessity in the sense of "would have been no matter what" or "couldn't have been otherwise" were largely overlooked until the later sixties and early seventies – not that everyone pays sufficient attention to such distinctions even today. That is to say, we were in possession of a model theory for modal logic well before most modal logicians had an unambiguous and unconfused understanding of what the modalities mean, or even any firm guarantee that the modalities do mean something. This fact alone should warn us that it is one thing to have a theory of models, and another to have a theory of meaning. The following argument is utterly invalid:

These sentences have models.
These sentences have a semantics.
These sentences have a meaning.

Philosophically speaking, the model theory for modal logic has much less *direct* value than does the parallel model theory for temporal logic. This is because different theories about time do naturally present themselves as theories about the relative-futurity relation on stages of the world, whereas different conceptions of modality do *not* naturally present themselves as theories about the relative-possibility relation on states of the world. It is very useful to know that a certain class of modal models corresponds to a certain axiomatic system, *if* there is reason to be interested in that axiom system. But philosophically speaking, model theory is of little direct use in establishing that a given axiom system is appropriate for a given conception of modality. Indeed, though we have had the model theory for getting on towards a half-century, the correct axiom system has been convincingly determined only for a couple of conceptions of necessity: truth by virtue of logical form, and provability in a given formal system.

I have addressed these topics elsewhere, but let me add just a word here about provability logic. Here we have a clear understanding or intended interpretation of what the box and diamond are to mean, and hence a clear understanding or intended interpretation of what it is for a sentence involving boxes and diamonds to be a law of logic, true in all instances – essentially, for all substitutions of sentences of first-order arithmetic for the sentence letters *p*, *q*, *r*, and so on. We have a proof that a certain axiomatic system, **GL**, is sound for this intended interpretation. We have also, thanks to Segerberg, a proof that **GL** is both sound and complete for a certain class of modal models. But for a long time this is *all* that we had, and we did not have what we want from a theory of models, namely, an assurance that truth in all models corresponds to truth in all instances, according to our intended interpretation. This is something like the Kreisel gap noted earlier in connection with first-order logic, but very much more serious.

But the serious lack was eventually supplied by the genius of Solovay. As it happened, he made use of Segerberg's earlier result. That is to say, rather than ignore the modal models altogether and directly establish that if a sentence is consistent with **GL** then it is true in some arithmetical instance, thus establishing the completeness of **GL** for the intended interpretation, he showed how from a model of the appropriate class of a sentence to produce sentences of first-order arithmetic that, substituted for the sentence letters, would result in an instance that is true according to the intended interpretation, thus showing that truth in all models does correspond to truth in all instances, and thus *indirectly* establishing the completeness of **GL** for the intended interpretation, given Segerberg's result about its completeness for models of the appropriate class. The model theory is useful, but it is useful only as an auxiliary. Segerberg's contribution did not go to waste, but Solovay's contribution was crucial. For a full exposition of these matters, and references to the further literature, see Boolos (1993).

There are other examples of the same phenomenon, where the model theory plays only an auxiliary role. One pertains to intuitionistic logic, with Tarski and Kripke in the role of Segerberg and Kreisel in the role of Solovay. For an exposition see Burgess (1981a). What reflection on these examples should make clear is that merely possessing a model-theoretic characterization of a given axiomatic system of modal logic does not suffice to tell us that the system captures (that is, is sound and complete for) any interesting interpretation or concept of necessity. Provision of a purely formal "semantics" for an axiomatic system that had been without one – for instance, the provision by Fine and by Meyer of a purely formal "semantics" for the system **R** of relevance/relevant logic – does little, and by itself does nothing towards establishing the coherence and intelligibility of any underlying motivating ideas or intended intuitive interpretation. This last is one case where there certainly was premature celebration, provoking a paper, "When is a semantics not a semantics?" by B. J. Copeland (1979), that is still well worth reading to dispel confusions. So Tarski's using the word "semantics" in connection with his truth definition has opened the door to confusion in the area of modal and more generally intensional or philosophical logic, which people like Copeland and myself have then had to come along after and try to straighten out.

6 FROM TARSKI TO DAVIDSONIANISM

Now one who opens a door that should have been left closed cannot be held responsible for every mischievous thing that then walks through it, and Tarski was himself the first victim of mischief resulting from his original offense. I think he was naughty, and should have been made to stand in a corner by linguists; but he actually suffered something considerably worse: he was stood on his head by Donald Davidson and his disciples. For Tarski's usage has not only tended to encourage the modal muddles I have been bemoaning, but also seems to me to have been in part responsible for the insufficiently critical reception of the truth-conditional theory of meaning; and this theory is diametrically opposed to Tarski's own views on the meaning of truth and to his views on the paradoxes with which we began. Now while I hope it will be universally agreed that the modal muddles Tarski's usage may have encouraged are bad things, this certainly will not be agreed about the acceptance of a truth-conditional theory of meaning or the rejection of Tarski's view of the paradoxes. The most I can hope will be agreed is that if Tarski's usage contributed to the formation of a predisposition to accept the one and reject the other that was independent of the real merits of the case, then *that* development was a bad thing.

In any case, besides Tarski's usage of "semantics" there is another feature of Tarski's definition of truth, or rather, of the Tarski–Vaught definition of the model, that may have been operative. To begin with, it may be noted that once we make the transition to the Tarski–Vaught notion of truth in a model, we soon find ourselves departing from Tarski's original notion that truth for a given language is to be defined in a metalanguage containing that language.

For the languages for which truth in a model are being defined are formal languages, while the papers on model theory containing the definitions are, like other mathematical papers, invariably written in (a mathematicians' dialect of) a natural language, nowadays usually English. Thus instead of

(7) $\mathfrak{M} \models (A \wedge B) \leftrightarrow (\mathfrak{M} \models A \wedge \mathfrak{M} \models B)$

(8) $\mathfrak{M} \models (A \text{ and } B) \text{ iff } (\mathfrak{M} \models A \text{ and } \mathfrak{M} \models B)$

we find

(9) $\mathfrak{M} \models (A \wedge B)$ iff ($\mathfrak{M} \models A$ and $\mathfrak{M} \models B$).

And whereas (7) and (8) do not on the face of it look as if they could be telling one anything about the meaning of "∧" or "and" that one did not already know – after all if one doesn't know the meaning of "∧" already one is not going to understand the right-hand side of (7), and if one does not know the meaning of "and" already one is not going to understand the right-hand side of (8) – by contrast (9) does rather look as if it were telling us something about the meaning of "∧," given that we already know the meaning of "and."

It is tempting to think of the status of (9) as being something like the status of the following:

(10) (Γ και Δ) is true iff Γ is true and Δ is true.

And (10) does tell us something – I do not say *everything*, but I do say *something* – about the meaning of the Greek word "και." But notice that (10) tells us something about the meaning of "και" only because we already know the meaning of "and" *and because we already know the meaning of* "true," at least to the extent of knowing Convention T. So there is a significant difference between (10) and (9), since the latter involves the symbol "⊨" rather than the English word "true."

If we consider the clauses in a Tarski–Vaught style definition, thus:

(11a) $\mathfrak{M} \models \sim A$ iff not $\mathfrak{M} \models A$
(11b) $\mathfrak{M} \models (A \wedge B)$ iff ($\mathfrak{M} \models A$ and $\mathfrak{M} \models B$)
(11c) $\mathfrak{M} \models (A \vee B)$ iff ($\mathfrak{M} \models A$ or $\mathfrak{M} \models B$)

we cannot take these to be telling us *both* what the double turnstile means *and* what the caret, wedge, tilde, and so on mean. For there are too many unknowns and not enough equations. The intention is that the double turnstile is to be read as "true," the caret as "and," the wedge as "or"; but the biconditionals would be equally appropriate if instead the double turnstile were read as "false" and the caret as "or" and the wedge as "and." If a student with no previous knowledge of these matters is told that the caret, wedge, and so forth are to be read as "and," "or," and so forth, then the student may be able to figure out that the double turnstile might be read as "true"; inversely, if the student is told that the double turnstile is to be read as "true," then the student may be able to figure out that the caret might be read as "and," the wedge as "or," and so forth. But the student has to be given some clue or other.

Now, once the model theorist is past the student stage, and knows that the double turnstile is customarily to be read as "true," the model theorist does learn something about how an addition to the usual list of logical symbols might be read, if the colleague proposing the addition indicates how the usual definition of double turnstile is to be extended to the extended language. For instance, a clause

(12) $\mathfrak{M} \models (A \downarrow B)$ iff not ($\mathfrak{M} \models A$ and $\mathfrak{M} \models B$)

would suggest the reading "not both" for the down arrow. And to give a perhaps more realistic example, in papers on generalized quantifiers the *same* symbol Q gets used over and over, with *different* clauses in the definition of double turnstile in different papers, suggesting different readings in different papers, among them "there exist infinitely many," "there exist uncountably many," "most," and others. The analogue of (12) in a given paper in such cases tells one something about what the symbol Q is being used to mean in that given paper. But our ability to guess the readings is entirely dependent on our familiarity with the meanings of "infinitely many" and "uncountably many" and "most," and on our familiarity with what "true" means, at least to the extent of being familiar with Convention T, and with the custom of reading the double turnstile as "true."

And in any case, we are not really being told that "$QxA(x)$" in a given paper means "there exist infinitely many x such that $A(x)$," but only that it means something that is true if and only if there exist infinitely many x such that $A(x)$, a condition fulfilled not only by "there exist infinitely many x such that $A(x)$" but also by "it is not the case that for all but finitely many x it is not the case that $A(x)$," and by infinitely many other alternatives. After all, merely to be told that some sentence of a foreign language we do not understand is true if and only if, say, snow is white, does not tell us what the sentence *means*. Taken together with our knowledge that snow *is* white, it does tell us that the sentence means something true. But that is all. Merely being given a list of items of the following kind, one for each sentence of the foreign language:

(13) "[FOREIGN SENTENCE]" is true iff [ENGLISH SENTENCE]

will not tell us what the sentences mean. If I tell you

(14a) "Το χιονι ειναι ασπρο" is true iff snow is white
(14b) "Το καρβουνο ειναι μαυρο" is true iff coal is black

I allow you, given the common knowledge that snow is white and coal is black, to infer that the two Greek sentences quoted are true; but I do not divulge the meaning of any Greek sentence.

Donald Davidson conjectured, however, in Davidson (1967) and sequels, that if we require some *finite* apparatus to generate *recursively*, with clauses like (10), a whole list of items of type (12) for every Greek sentence, and if we impose some suitable further restrictions, then in fact what appear on the right-hand side of each item on the list will have to be English sentences with the same meaning as the Greek sentence on the left-hand side. Or rather – since he was writing during the era of Quinean suspicion about meaning – the Greek and English sentences will have to be close enough in meaning, according to our intuitive, pre-theoretic understanding of "meaning," that appearing on opposite sides of a list generated in the manner indicated can serve as a workable substitute for the intuitive, pre-theoretic, but to some suspect, understanding of sameness of "meaning." In short, while in order to learn the meaning of Greek sentences and the words of which they are composed it is not enough to be *told* things like (14a,b), Davidson conjectured that, in order to do so it may be enough to be told, or to come to know, such things *in the right way*.

Davidson's conjecture that theories of truth can in this sense serve as theories of meaning eventually gave rise to what I will call *Davidsonianism*, without intending to imply that Davidson himself fully subscribed to it, namely, the truth-conditional theory of meaning. Formally, indeed, the step from Davidson's theory about how to tell the meanings of foreign sentences to speakers who already know the meaning of English sentences to the Davidsonian theory that knowledge of the truth conditions of English sentences is what knowledge of the meaning of English sentences consists in, can be a very short one. For the simplest version of Davidsonianism would simply be the analogue of what Davidson says about Greek and English applied to English and a hypothetical innate language of thought. Using the language of Descartes to represent the language of thought, coming to know the meaning of the English sentences "Snow is white," "Grass is green," and so on, amounts to coming in the right way to know the following:

(15a) "Snow is white" *est vraie ssi la neige est blanche*
(15b) "Grass is green" *est vraie ssi l'herbe est verte.*

Davidsonianism as such, it should be emphasized, is not committed to the language of thought hypothesis; I only say that Davidsonianism is immediate from the conjecture of Davidson if one accepts that hypothesis.

For some of us Davidsonianism seems, not least on account of its apparent assumption that truth is an innate idea, possession of which is a prerequisite for all language-learning, to be preposterous. For others, the

Davidsonian assumption is so much taken for granted that it is hardly recognized as a substantive assumption at all. My concern here will be not with the enormous question of the merits or demerits of the truth-conditional theory of meaning, but only with the extent to which Tarski deserves a share of the blame or credit for its becoming so widely held a belief among philosophers as it currently is.

The first thing that must be said on this head is, of course, that Tarski's responsibility is limited. Davidson himself is not fully responsible for his disciples' extrapolations from his conjectures, and Tarski is certainly not fully responsible for widespread sympathetic reception of those conjectures. It must be acknowledged, for one thing, that their sympathetic reception was surely in part due to the prestige of the name of Davidson as a result of quite other achievements. But then again, was it not in part due to the prestige of the name of Tarski, which Davidson so frequently invoked?

Well, even if so, it must be acknowledged, for another thing, that the invocation of Tarski's name was not entirely appropriate, since as Davidson, if not every one of his disciples, was aware, those conjectures amount to an inversion of Tarski. For they make what for Tarski were clauses in a definition of truth in terms of already understood notions like negation and conjunction and disjunction, into definitions of a kind of those operators, in terms of a notion of truth taken as primitive. We constantly find in the writings of Davidson and disciples mentions of a "Tarskian" theory of truth, where "counter-Tarskian" or "anti-Tarskian" would have been more accurate, if less likely to confer borrowed prestige on bold (which is to say doubtful) new conjectures. And Tarski, of course, is not responsible for this usage. But would the idea of invoking Tarski's name at all in connection with a theory of meaning have occurred to anyone, if Tarski had not himself attached to his theory a label ordinarily used for the theory of meaning, the label "semantics"?

7 THE "SEMANTIC" PARADOXES

Well, if so then one important consequence of Tarski's usage of "semantics" was to help popularize a theory with implications diametrically opposed to his own view on the "semantic" paradoxes, his so-called inconsistency theory of truth. This is the last point I wish to bring out, and it requires a little background. The natural anti-Davidsonian assumption is that understanding the meaning of a word consists, not in knowledge of how it contributes to the truth conditions of sentences in which it appears, but rather in internalization of rules for its use. For a "use" theorist, it would be nothing short of a miracle if we never internalized any rules having lurking inconsistencies, and the intractability of the liar and related paradoxes strongly suggests that in the case of "truth" the rules may be the obvious, inconsistent ones, permitting free passage back and forth between *p* and "it is true that *p*," in accordance with an absolutely unrestricted version of Convention T.

Now while Tarski neither explicitly endorses the "use" theory nor explicitly says that Convention T is all there is to the intuitive meaning of "true" – though it represents all *he* is able to understand about the pre-theoretic meaning of "true" – he does endorse one consequence of those views, namely, the consequence that the intuitive notion of truth is the inconsistent one. For Tarski, what is to be expected of a theory of truth is therefore not a *vindication* of the intuitive notion, but a restricted *replacement* for it: a serviceable substitute applicable not to all languages, but only to languages of a certain comparatively simple structure. This may be the view of most of us who have worked on the technical comparative study of the various theories of truth that emerged in the wake of Kripke's famous "outline" – or if not the majority view, anyhow the plurality view, having more adherents than any particular one of the more positive theories. But among philosophers generally it is distinctly a minority view, subscribed to by my teacher Charles Chihara, my student John Barker, and (*not* as the result of any influence of mine) my son Alexi Burgess, and by very few others. Indeed it seems that many if not most philosophers are so violently prejudiced against this view that they do not even wish to contemplate what *would* be an appropriate response if the theory *were* accepted.

While there may be various other reasons for this prejudice, surely one of the most important is that the inconsistency theory of truth is incompatible with the truth-conditional theory of meaning. Insofar as his own usage of "semantic" tended to encourage the view that "Tarskian" truth conditions are central to meaning, Tarski has himself ironically helped to create a prejudice against his own views on the paradoxes and their lessons. (Here as in so many other cases the malefactor has been himself the chief victim of his malefaction.) And though the question of merits or demerits of the inconsistency theory of truth, like the issue of the merits or demerits of the truth-conditional theory of meaning, of which issue it is but one aspect, is too large a question to be gone into here, I hope it could at least be agreed that such questions ought to be examined without prejudice. Since nothing, perhaps, does more to encourage a bias in favor of the view that "Tarskian" truth conditions are central to meaning than the fact that Tarski himself calls the approach to truth involving such inductive clauses "semantic," we have here another reason, additional to the modal muddles mentioned earlier, for avoiding that usage. Let us therefore, in roaming the vast field Tarski opened up for us, not follow him in his one terminological trespass. Let us honor him for every aspect of what he called the "semantic" conception of truth *except* for his calling it that.

9
Which modal logic is the right one?

I THE QUESTION

Which if any of the many systems of modal logic in the literature is it whose theorems are all and only the right general laws of necessity? That depends on what kind of necessity is in question, so I should begin by making distinctions.

A first distinction that must be noted is between *metaphysical necessity* or *inevitability* – "what could not have been otherwise" – and *logical necessity* or *tautology* – "what it is self-contradictory to say is otherwise." The stock example to distinguish the two is this: "Water is a compound and not an element." Water could not have been anything other than what it is, a compound of hydrogen and oxygen; but there is no self-contradiction in saying, as was often said, that water is one of four elements along with earth and air and fire.

The logic of inevitability might be called *mood logic*, by analogy with tense logic. For the one aims to do for the distinction between the indicative "it is the case that . . ." and the subjunctive "it could have been the case that . . .," something like what the other does for the distinction between the present "it is the case that . . ." and the future "it will be the case that . . ." or the past "it was the case that . . ." The logic of tautology might be called *endometalogic*, since it attempts to treat within the object language notions that classical logic treats only in the metalanguage. However, it hardly deserves a name, since it immediately splits up into two subjects.

For a second distinction must be between two senses of tautology. On the one hand, there is *model-theoretic logical necessity* or *validity*, the non-existence of a falsifying interpretation, "being true by logical form alone." On the other hand, there is *proof-theoretic logical necessity* or *demonstrability*, the existence of a verifying derivation, "being recognizable as true by logical considerations alone." Likewise, there is a distinction between two notions of contradiction, model-theoretic unsatisfiability and proof-theoretic inconsistency, and between two notions of implication, model-theoretic consequence and proof-theoretic deducibility. There would be at least a conceptual distinction even if logic were understood narrowly as first-order logic, where the model-theoretic and proof-theoretic notions coincide in extension by the Gödel Completeness Theorem. There may be a difference in extension between them when logic is understood more broadly: for instance, if it is taken to include higher-order logic and the mathematics that goes with it.

Logicians often call model-theoretic and proof-theoretic necessity *semantic* and *syntactic* logical necessity. However, there is a conflict between this usage and the older usage of linguists on which, roughly speaking, "semantic" means "pertaining to meaning" and "syntactic" means "pertaining to grammar." There is a conflict between the two usages of "semantics," especially, because there is or may be a gap between mathematical modeling and intended meaning. In any case, shorter labels than "the logic of semantic logical necessity" and "the logic of syntactic logical necessity" would be useful. One might use *proplasmatic* logic and *apodictic* logic, from the Greek for model and proof. But it may be more suggestive to use *validity logic* and *demonstrability logic*, by analogy with *provability logic*. The analogy between provability logic and demonstrability logic is especially close, the one being concerned with what a theory can prove, the other with what we can demonstrate, the "can" in each pertaining to ability in principle, regardless of practical limitations.

The question which is the right system of tense logic is not one for the logician: the logician can indicate how this or that or the other system corresponds to this or that or the other theory of the nature of time, but which is the right theory of the nature of time is a question for the physicist. Similarly, the question which is the right system of mood logic would seem to be one not for the logician, but for the metaphysician. By contrast, the question which is the right system of validity or demonstrability logic cannot be passed off by logic to some other discipline.

The question which is the right validity logic has been answered at the sentential level, which is the only level that will be considered here: it is the system known as S5. This result is essentially established already in Carnap (1946).

The question which is the right demonstrability logic, again at the sentential level, goes back to the earliest days of modern modal logic. For though the founder of the subject, C. I. Lewis, did not clearly distinguish among metaphysical, model-theoretic logical, and proof-theoretic logical modalities, still he did always write of necessitation as implication, and did often write of implication as deducibility, so that it is reasonable to conclude that by necessity he primarily meant tautology, by which in turn he primarily meant demonstrability. No one today, however, takes seriously his suggestion that the right logic for this notion might be the feeble **S1** or the bizarre **S3**. To the extent that there is any consensus or plurality view among logicians today, I take the view to be that the right demonstrability logic is **S4**. (Even in "relevance" or "relevant" logic, where **S4** cannot be literally accepted, since the classical sentential logic it is based on is rejected, still it seems to be a consensus or plurality view that the right logic should be "**S4**-like.") The *locus classicus* for such a view is a paper from the proceedings of a famous 1962 Helsinki conference on modal logic (Halldén 1963).

While the argument for the soundness of **S4** as a demonstrability logic given there seems as compelling as an "informally rigorous" argument can be, there is no real argument for completeness, which remains an open question. It therefore remains conceivable that the right logic is something stronger than **S4**: that it is something intermediate between **S4** and **S5**, such as **S4.2** or **S4.3**; or that it is something stronger than **S4** but incomparable with **S5**, such as the logic called **Grz** after Grzegorczyk (1967) and the logic that ought to be called **McK** after McKinsey (1945). (In the literature it has heretofore been misleadingly called **S4.1**, though it is not intermediate between **S4** and **S5**.)

The issues are sufficiently illustrated by the cases of the distinctive axioms of **S4.2** and of **McK**, which are equivalent respectively to $\sim(\Box\sim\Box p \wedge \Box\sim\Box\sim p)$, the principle that "nothing is both demonstrably not demonstrable true and demonstrably not demonstrably false," and to $\Box\sim\Box p \vee \Box\sim\Box\sim p$, the principle that "everything is either demonstrably not demonstrably true or demonstrably not demonstrably false." Halldén rightly says of the latter – what he could also have said of the former – that it is not an intuitively plausible principle when the box $\Box$ is meant as demonstrability. But to say this is to do something less than to give an "informally rigorous" argument for the claim that either principle outright fails as a general law, let alone for the claim that any principle not a theorem of **S4** does so.

The question which is the right provability logic has been answered, and though results are often stated for a single-theory, classical first-order arithmetic, many hold for all true theories satisfying certain minimum requirements of strength. Actually, one must distinguish the question of which logic gives all and only those principles about provability all whose instances are provable by the theory in question from the question which gives all and only those principles that are valid (or demonstrable by us).

The answer to the former question is given by a system **GL**, and to the latter question by a system **GLS**. Both differ from the Lewis systems **S4** and **S5** by lacking the law $\Box(\Box p \rightarrow p)$. The failure of this law is, roughly speaking, the content of the Gödel Incompleteness Theorems. The standard reference is of course Boolos (1993).

Below, in §2 I will recall the case for the soundness of **S5** as a validity logic and of **S4** as a demonstrability logic. In §3 I will recall the Carnapian case for the completeness of **S5**. In §4 I will indicate the minimal requirements of strength that are assumed in provability logic, and that I will be assuming in demonstrability logic also, and attempt to clarify the relationship between the two logics. In §5 I will present a case against **McK** as a demonstrability logic; and it will generalize to a case against any system not contained in **S5**, such as **Grz**. Finally, in §6 I will present a case against **S4.2**; and this will of course also constitute a case against any stronger system, such as **S4.3**. But the case of weaker systems intermediate between **S4** and **S5** will be left open, and with it the general question.

2 SOUNDNESS

A key consequence of the step of treating modality in the object language, treating $\Box$ as a one-place connective on a par with $\sim$, is that *iterated* modalities, modalities embedded inside modalities, as in $\Box\sim\Box p$, are allowed. By contrast, when "valid" is expressed only by a word of the metalanguage, applicable only to formulas of the object language, there can be no question of iterations like "it is valid that it is not valid that . . ." All the modal systems most commonly considered in the literature agree with classical meta-logic, in the sense that where classical meta-logic has a law, for example "a valid conclusion is a consequence of any premise," these systems will have a corresponding law, in the example $\Box p \rightarrow (q \Rightarrow p)$. Agreeing as they all do with classical logic, these systems agree with each other for formulas without iterated modalities. What distinguishes **S5** is that it has laws that make every iterated formula more or less trivially equivalent to an uniterated formula.

The first step in establishing **S5** as the right logic of validity is to establish the *soundness* of **S5** as a validity logic: to establish that every theorem of **S5** is correct as a general law about validity, or what comes to the same thing, that every axiom of **S5** is thus correct, and that every rule of **S5** preserves such correctness. This is completely unproblematic for the non-modal axioms and rules, which are simply those of the sentential component of classical, non-modal sentential logic. Moreover, though **S5** is usually formulated with a specifically modal rule allowing $\Box A$ to be taken as a theorem whenever A is a theorem, this rule can be dispensed with in favor of adding $\Box A$ as an axiom whenever A is an axiom of the usual formulation, which is to say, whenever A is either a classical, non-modal axiom or one of the following modal axioms:

(1) $\Box p \to p$
(2) $\Box(p \to q) \to (\Box p \to \Box q)$
(3) $\Box p \to \Box\Box p$
(4) $\sim\Box p \to \Box\sim\Box p$.

Again for the classical, non-modal axioms this is completely unproblematic, while in making a case for – which is to say, in demonstrating – any one of (1)–(4), one will at the same time be making a case for its demonstrability, and a fortiori for its validity.

Thus the problem of establishing the soundness of S5 for validity logic reduces to that of establishing the correctness of (1)–(4), and similarly the problem of establishing the soundness of S4 for demonstrability logic reduces to that of establishing the correctness of (1)–(3). Indeed, for (1)–(3) correctness for the box as validity and for the box as demonstrability can be established by more or less parallel arguments. Consider (1), for example. The arguments are simply the parallel ones that whatever is true by logical form alone must be true, and that whatever can be recognized to be true by logical considerations alone must be true. But indeed, I need not enlarge on the case for (1)–(3), which is adequately made by Halldén.

It remains to consider the distinctive axiom (4), which of course is being proposed only as a general law of validity, not of demonstrability. Here the main point is just as follows. Consider any particular instance of (4):

(5) If it is not true by logical form alone that π, then it is true by logical form alone that it is not true by logical form alone that π.

Suppose that the antecedent of (5) is true, which is to say that the following is false:

(6) It is true by logical form alone that π.

Since (6) is false, there must be some ψ of the same logical form as π that is false. Now consider anything else of the same logical form as (6). It will look like the following, wherein ρ has the same logical form as π:

(7) It is true by logical form alone that ρ.

But then ψ also has the same logical form as ρ, and since ψ is false, (7) is false. In other words, anything of the same logical form as (6) is false, and hence the following is true:

(8) It is true by logical form alone that it is not true by logical form alone that π.

Thus the consequent of (5) is true, as required.

3 COMPLETENESS FOR VALIDITY LOGIC

In the case of provability logic, as expounded in Boolos (1993), and of intuitionistic logic, as expounded in Burgess (1981a), once the candidate logic S has been identified, the argument that it is the right one consists of three parts: soundness, *formal* completeness, and *material* completeness. That is, it is shown that every theorem of S is acceptable as a general law under the intended interpretation; a class Σ of mathematical models is identified and it is shown that (every theorem of S and) no non-theorem of S comes out true in all models of class Σ; and it is shown that any formula that comes out untrue in some model of class Σ is unacceptable as a general law under the intended interpretation. This last step, bridging the gap between "semantics" in the logicians' sense and "semantics" in something more like the linguists' sense, is due in the case of provability logic to R. M. Solovay, and in the case of intuitionistic logic to Georg Kreisel, who coined the phrase "informal rigor" in this connection. In both cases, the last step is the most difficult. The situation is rather similar in the case of validity logic.

Beginning with formal completeness for the case of validity logic (soundness having already been discussed), the kind of models now standard in modal logic first became widely known through another talk at the 1962 Helsinki conference, this one by Saul Kripke; another kind of model near to those standardly used became widely known through yet another talk at the same conference, this one by Jaakko Hintikka. A *frame* model M, as in Kripke (1963), consists of two parts, a *frame* and a *valuation*. The frame consists of a set W of indices, a two-place relation R on it, and a designated member w_0 of it. A valuation V is a specification for each x in W and each sentence letter p, q, r, ... of whether or not the sentence letter counts as true in that index. The notion of truth in an index is extended to compound formulas by recursion:

$\sim A$ is true at x if and only if A is not true at x.
$A \wedge B$ is true at x if and only if A is true at x and B is true at x.
$\Box A$ is true at x if and only if A is true at y for every y in W such that Rxy.

It is permitted to have two indices x and y at which exactly the same set of formulas are true, and such *duplication* is often important. A formula counts as holding in M if it is true at w_0, and as being valid in a class K of frames if it holds in every model whose frame is in that class.

The proposal in Hintikka (1963) is less purely "semantic" or model-theoretic: it is still "syntactic" or proof-theoretic in that, while it has a relation R, what this relation relates are not abstract indices, but sets of formulas, and so one does not have duplication. But as it happens, in connection with the system **S5**, differences between the approaches of Kripke and Hintikka, such as permitting or forbidding duplication, are unimportant; and so for that matter is the main similarity between the two approaches: the presence of a relation R. For while the theorems of **S5** can be characterized as the formulas valid for the class of reflexive, transitive, and symmetric frames, this characterization reduces, by a series of steps too familiar to bear repetition here, to a much simpler one.

Consider, for any k, the formulas involving only the sentential variables or atomic formulas $p_1, \ldots, p_k$. Then for such formulas, a model may be taken to consist simply of a non-empty subset W of the set of rows of the truth table $p_1, \ldots, p_k$, with one such row w_0 designated. The notion of truth at a row x in the truth table is defined for compound formulas by a recursion in which the first two clauses are exactly the same as above, while the third reads as follows:

$\Box A$ is true at x if and only if A is true at y for every y in W.

A formula counts as holding in such a model M if it is true at w_0. The theorems of **S5** may be characterized as the formulas that hold in all such models.

Turning to material completeness for the case of validity logic, it may be well to begin by considering an objection to the modeling just described that has been independently advanced by several writers. One of them put it as follows:

> What is needed for logical necessity of a sentence p in a world w_0 is more than its truth in each one of some arbitrarily selected set of alternatives to w_0. What is needed is its truth in each *logically possible* world. However, in Kripke semantics it is not required that all such worlds are among the alternatives to a given one.

It is then suggested that one should adopt not the standard model theory, or the simplification thereof described above, but rather a deviant model theory, which after simplification amounts to just this, that the only model admitted is the one consisting of *all* rows of the truth table.

There is a fallacy or confusion here. What is wanted is that the technical notion of coming out true in all models should correspond to the intuitive notion of coming out true under all interpretations, or all substitutions of specific $\pi_1, \ldots, \pi_k$ for the variables $p_1, \ldots, p_k$. Since, for instance, among all the many substitutions available there are ones in which the π_1 substituted for p_1 *is the same as* the π_2 substituted for p_2, so that it is impossible for π_1 and π_2 to have different truth values, there must correspondingly be among the models one available where the only rows of the truth table present are those for which the value given to p_1 *is the same as* the value given to p_2.

The confusion in the objection becomes apparent when one notes that in the deviant model theory suggested, $\sim\Box\sim(p_1 \wedge \sim p_2)$ counts as valid, whereas of course $\sim\Box\sim(p_1 \wedge \sim p_1)$ does not, so that the standard rule of substitution fails. But the rule of substitution must hold so long as one adheres to the standard conception of the role of the variables $p_1, \ldots, p_k$, according to which arbitrarily selected $\pi_1, \ldots, \pi_k$ may be substituted for them. Indeed, the deviant model theory corresponds to a deviant conception on which *independent* $\pi_1, \ldots, \pi_k$ *must be substituted for distinct* $p_1, \ldots, p_k$. The confusion is worse confounded when it is suggested that the difference between the standard and deviant model theories somehow corresponds to a difference between non-logical and logical notions of necessity. For what is at issue is, to repeat, differences in conceptions of the role of variables, not in conceptions of the nature of necessity.

Yet, confused as it is, the objection does serve to call attention to an important question. Each substitution of specific $\pi_1, \ldots, \pi_k$ for $p_1, \ldots, p_k$ determines a non-empty set of rows of the truth table, consisting of all and only those rows x such that it is not impossible by the logical forms of the π_i alone for them to have the truth values x assigns to the corresponding p_i. The question is, is it the case that for any arbitrarily selected non-empty subset W of the set of rows of the truth table, there are specific statements $\pi_1, \ldots, \pi_k$ that determine, in the manner just described, exactly that subset? In other words, if a formula is not a theorem of **S5**, and therefore fails in some standard model, is there some specific instance in which it fails? An affirmative answer to this question is precisely what is needed to establish the material completeness of **S5** as validity logic.

It is a reasonable assumption, and one presumably made by the critics alluded to, that there exist indefinitely many $\alpha, \beta, \gamma, \ldots$ that are independent in the sense that any conjunction of some of them with the negations of the rest of them is possible, in the relevant sense of possibility. For instance, if α, β, γ, . . . are of simple subject–predicate form with distinct subjects and predicates in each, they will be thus independent. Given this assumption, an affirmative answer to the foregoing question is forthcoming. As this result has in effect already been expounded several times in the literature, in Carnap (1946), Makinson (1966), and S. K. Thomason (1973), there should be no need for me to do more than give an illustrative example here. Indeed, a simple one, involving just three sentence letters p, q, r, should suffice.

Consider the set W containing just the three rows in which two of p, q, r are true and the other false. Call the one where r alone is false x, the one where q alone is false y, and the one where p alone is false z. What is to be established is that given independent α, β, . . . , there are truth-functional compounds π, ψ, ρ thereof that might be substituted for p, q, r, for which the three rows indicated represent all and only the combinations of truth values that are not false by logical form alone.

To find the required compounds, one first finds three auxiliary compounds ξ, υ, ζ, that are pairwise exclusive and jointly exhaustive, meaning that the conjunction of any two must be false by logical form alone, while the disjunction of all three must be true by logical form alone. Setting $\xi = \alpha$ and $\upsilon = \sim\alpha \wedge \beta$ and $\zeta = \sim\alpha \wedge \sim\beta$ will do. One next lets the auxiliaries ξ, υ, ζ correspond to the rows x, y, z, and takes as the substitute for a given one of p, q, r the disjunction of the auxiliaries corresponding to the rows in which it is true. Thus the substitute π for p should be $\xi \vee \upsilon$ or $\alpha \vee (\sim\alpha \wedge \beta)$, which simplifies to $\alpha \vee \beta$. It can be worked out that the substitutes ψ and ρ for q and r simplify to $\alpha \vee \sim\beta$ and $\sim\alpha$, respectively. And it can then be worked out that exactly two of the three, $\alpha \vee \beta$ and $\alpha \vee \sim\beta$ and $\sim\alpha$, must be true, and that given the independence of α and β it may be any two of the three, as required.

Before leaving the topic of validity logic it may be mentioned that the fact that S5 is indeed the right logic can be confirmed in a different way. After soundness is established in order to show that no stronger system than S5 is acceptable, one would appeal to the result of Scroggs (1951), according to which the only extensions of S5 are finitely-many-valued logics. One would then argue that no finitely-many-valued logic can be correct for semantic logical necessity (given the same reasonable assumption as above, that there are indefinitely distinct independent statements).

4 DEMONSTRABILITY AND PROVABILITY

A word must be said to clarify the relationship between demonstrability and provability logics, and to dispel a puzzle about that relationship.

The minimal assumptions of strength needed for provability logic are three. They may be formulated either as assumptions on the notion of proof for the theory, or as assumptions on the set of theorems provable. On the formulation in terms of proofs, the first assumption would be that whether something is or is not a proof in the theory is decidable, which by Church's Thesis implies that the relation of proof-in-the-theory to theorem proved is recursive. The second assumption would be that the rules of classical first-order logic may be used in proofs in the theory. The third assumption would be that certain basic, finite, combinatorial modes of reasoning – whose exact scope need not be gone into here, except to say that, since we want to get the Second Incompleteness Theorem, the scope needs to be somewhat wider than it would need to be if we only wanted to get the First Incompleteness Theorem – may be used in proofs in the theory.

On the formulation in terms of theorems, it would first be assumed that the set of theorems provable is recursively enumerable. It would second be assumed that the set of theorems provable is closed under the rules of classical first-order logic. And it would third be assumed that the set of theorems provable includes certain basic, finite, combinatorial results. Clearly, the list of assumptions on the theory stated earlier yields the list of assumptions on the set of theorems just stated. And conversely, by Craig's Lemma, any set of conclusions satisfying this latter list of assumptions coincides with the set of conclusions provable in some theory satisfying the former list of assumptions.

In demonstrability logic, at least as I will be considering it here, it is to be assumed that whether something constitutes a demonstration of a given conclusion is decidable, that the rules of classical first-order logic may be used in demonstrations, and that certain basic, finite, combinatorial modes of reasoning may be used in demonstrations. By what has already been said, it follows that the set of conclusions we can demonstrate coincides with the set of conclusions that can be proved in some theory of the kind to which provability logic applies. And yet, provability logic and demonstrability logic are supposed to be different, in that by what has been said in earlier sections, (1) below is false, while (2) below is true:

(1) It can be proved in such-and-such a theory T that if something can be proved in such-and-such a theory T, then its negation cannot also be.

(2) It can be demonstrated by us that if something can be demonstrated by us, then its negation cannot also be.

What may be puzzling is how it can be that (1) fails while (2) holds and as already indicated the above-stated assumptions commit one to the truth of something of the following form:

(3) What can be demonstrated by us coincides with what can be proved in such-and-such a theory T.

Indeed, a notorious objection to (3) above, associated with the names of J. R. Lucas and Roger Penrose, claims that it, together with the true (2) above, yields the false (1) above.

The solution to the puzzle is to point out that this objection commits the fallacy of assuming that co-extensive terms, such as "what we can demonstrate" and "what such-and-such a theory T can prove" can be substituted without change of truth value everywhere, even in intensional contexts, such as "we can demonstrate that . . . " or "such-and-such a theory T can prove that . . ." To get (1) above from (2) above one would need something stronger than (3) above, namely the following:

(4) We can demonstrate (3).

The solution of the puzzle is that (3) does not yield (4). (By analogy, those familiar with the work of Solomon Feferman and William Tait on the standpoints of the predicativists and the finitists, thinkers who owing to their philosophical prejudices cannot demonstrate all that we can, will recall that what is demonstrable from the standpoint of one or the other of these 'isms can indeed be exactly captured by a theory, though the 'ists themselves cannot recognize as much.)

The fallacy should become obvious on comparing (1)–(3) above with the following:

(1′) The only man with such-and-such a number N of hairs on his head knows that the only man with such-and-such a number N of hairs on his head is gray-haired.

(2′) I know that I am gray-haired.

(3′) I am the only man with such-and-such a number N of hairs on his head.

Clearly (3′) above by itself does not, with (2′) above, yield (1′) above. Rather, one would need the following stronger assumption:

(4′) I know (3′).

But (3′) above does not yield (4′) above.

5 AGAINST McK

Beginning with formal completeness for the case of demonstrability logic, soundness having already been discussed, the theorems of S4 can be characterized as the formulas valid for the class of reflexive and transitive frames; and equally, they can be characterized as the formulas valid for the class of *finite* reflexive and transitive frames, a deeper result implying the decidability of the logic. A historical fact is worth mentioning, that the result just stated follows *immediately* from two results already in the literature two decades before the famous Helsinki conference. One of these, from McKinsey (1941), characterizes the theorems of S4 in terms of a class of finite models of a different kind, based not on frame structures but on algebraic structures of a certain kind. The other of these, from Birkhoff (1937), connects finite algebraic structures of the kind in question with finite reflexive and transitive frame structures. (The former paper makes no mention of frames, and the latter no mention of modal logic.) This history is worth mentioning among other reasons because the older algebraic modeling involved, which is sometimes not taught to students of the subject today, still has its uses even after the development of frame models, and I will be citing an instance later.

Turning to material completeness, no decisive results have yet been obtained, and my aim will only be to present, case by case, some partial results. To begin with, it is not only a reasonable assumption, as already said in an earlier section, that there exist indefinitely many α, β, γ, ... that are independent, but also a reasonable assumption that there exist indefinitely many α, β, γ, ... that are *demonstrably* independent. For instance, if α, β, γ, ... are *recognizably* of simple subject–predicate form with distinct subjects and predicates in each, they will be thus demonstrably independent. It follows that if *D* is any compound formed by negation and conjunction or disjunction from *p*, *q*, *r*, ... such that *D* is a not a theorem of classical sentential logic, or in other words, such that *D* comes out false in some row of the pertinent truth table, then the result Δ of substituting α, β, γ, ... for *p*, *q*, *r*, ... in *D* will be demonstrably not demonstrable, or demonstrably *indemonstrable*. Similarly, if *D* is such that ~*D* is not a theorem of classical sentential logic, or in other words, such that *D* comes out true in some row of the pertinent truth table, then Δ will be demonstrably not demonstrably false, or demonstrably *irrefutable*. Thus from any *D* such that neither it nor its negation is a theorem of classical sentential logic, we get a counterexample Δ to the McK axiom that nothing is both demonstrably indemonstrable and demonstrably irrefutable.

This argument can be generalized to apply to *any* axiom that is not a theorem of S5.

Perhaps the easiest route to a generalization is to draw on the work of Slupecki and Bryll (1973). They pursue the old idea of the Polish school that a logical system should have in addition to its axioms and rules of the ordinary kind, its axioms and rules of *acceptance*, indicating that certain formulas are acceptable as general laws, some axioms and rules of an opposite kind, axioms and rules of *rejection*, indicating that certain formulas are *un*acceptable as general laws. Just as a formula P is a theorem of the system, in symbols $\vdash P$, if there is a sequence of steps, each an axiom of acceptance or following from earlier ones by a rule of acceptance ending in P, so the goal would be to have for each formula Q that is not a theorem of the system, in symbols $\dashv Q$, a sequence of steps involving axioms and rules of rejection ending in Q.

For classical logic there would be the axiomatic rejection of the constant false $\dashv$ and $\perp$ rules of rejection that are the reverse of the usual classical rules of acceptance: if $\dashv P'$ where P' is a substitution instance of P, then $\dashv P$, and if $\dashv Q$ and $\dashv P \supset Q$, then $\dashv P$. For any modal logic there would be also the rule of rejection that is the reverse of the usual modal rule of acceptance: if $\dashv \Box P$ then $\dashv P$. For each particular modal system additional rules of rejection would be needed. For S5 Slupecki and Bryll show that just one additional rule suffices:

$$\text{if } \dashv P \rightarrow Q_1 \text{ and} \ldots \text{and } \dashv P \rightarrow Q_N \text{ then } \dashv \Box P \rightarrow (\Box Q_1 \vee \ldots \vee \Box Q_N)$$

where P and the Q_i involve no modalities.

In order to show that any non-theorem of S5 should be rejected as a general law of demonstrability, it will suffice therefore to argue that the above rule of rejection is acceptable for demonstrability. And indeed, if the $P \rightarrow Q_i$ are unacceptable as general laws, there must for each be a row x_i of the truth table for the variables $p, q, \ldots$ involved on which P comes out true and Q_i comes out false. But then by Carnap's result there are specific π, $\psi, \ldots$ that could be substituted for the variables $p, q, \ldots$ in P and the Q_i to give sentences Π and Ψ_i, such that the x_i represent all and only the possible combinations of truth values for the π, $\psi, \ldots$ It follows that Π is an instance of a theorem of classical sentential logic, hence demonstrable, while each Ψ_i is demonstrably indemonstrable by the considerations of the preceding paragraph. Thus the following:

$$\Box\Pi \rightarrow (\Box\Psi_1 \vee \ldots \vee \Box\Psi_N)$$

fails, and the formula of which it is an instance, namely the following:

$\Box P \rightarrow (\Box Q_1 \vee \ldots \vee \Box Q_N)$

is not acceptable as a general law, as required.

6 AGAINST S4.2

The S4.2 principle says that everything is either demonstrably indemonstrable, or demonstrably irrefutable. An argument against the acceptability of this principle as a general law can be given.

In addition to assumptions about demonstrability listed in earlier sections, including the assumption that there are indefinitely many instances recognizable as being of simple subject–predicate form, I need the reasonable assumption that there is an instance recognizable as being of simple subject–verb–object form, with no further pertinent logical structure. That is, I assume there is a two-place predicate Φ that is recognizably a two-place predicate with no further pertinent logical structure, so that it is recognizable that all its pertinent logical structure is represented when it is represented by a simple two-place predicate variable F. Given such an example, for any compound Π formed from Φ using negation, conjunction or disjunction, and universal or existential quantification, the logical form of Π will recognizably be represented by a formula P of classical first-order logic formed from F using $\sim$, $\wedge$ or $\vee$, and $\forall$ or $\exists$. Let Λ be the set of all such compounds Π, and L the set of the corresponding formulas P.

Now suppose a formula P in L fails in no model of the kind used in classical first-order logic. Then P is a theorem of classical first-order logic by the Gödel Completeness Theorem; and by the assumption that the rules of classical first-order logic may be used in demonstrations, the corresponding Π in Λ will be demonstrable, and demonstrably so. Then by general laws represented by theorems of S4, "it is demonstrable that Π" will be demonstrably irrefutable and not demonstrably indemonstrable.

Now suppose the formula P of L fails in some *finite* model of the kind used in classical first-order logic. Then basic, finite, combinatorial reasoning shows that it does so, and hence that it does not represent a correct general law; and by the assumption that basic, finite, combinatorial reasoning may be used in demonstrations, the corresponding Π in Λ will be indemonstrable, and demonstrably so. Then by general laws represented by theorems of S4, "it is demonstrable that Π" will be demonstrably indemonstrable and not demonstrably irrefutable.

For any formula P of L, let $\Psi(P)$ be "it is demonstrable that Π," where Π in Λ is the result of substituting Φ for F. Let X be the set of P such that $\Psi(P)$ is demonstrably irrefutable, and let Y be the set of P such that $\Psi(P)$ is demonstrably indemonstrable. What has been established so far is that if P has no counter-model, then P belongs to the difference set $X-Y$; while if P has a finite counter-model, then P belongs to the difference set $Y-X$. What the **S4.2** principle yields is that the union set $X \cup Y$ is all of L.

To complete the case against the **S4.2** principle, I must invoke the assumption that the set of demonstrable conclusions is recursively enumerable, from which it follows that the sets X and Y are also recursively enumerable. Then by the Reduction Theorem for recursively enumerable sets it follows that there are recursively enumerable sets X^* and Y^* satisfying the conditions that $X^* \subseteq X$ and $Y^* \subseteq Y$ and $X^* \cap Y^* = \emptyset$ and $X^* \cup Y^* = X \cup Y$. These conditions imply that $X - Y \subseteq X^*$ and $Y - X \subseteq Y^*$. What the **S4.2** principle yields is that X^* and Y^* are complements of each other in the recursive set L, which since both are recursively enumerable yields that both are recursive. What was established earlier yields that if P has no counter-model, then P belongs to X^*, while if P has a finite counter-model, then P belongs to Y^* and so does not belong to X^*. And now we have a contradiction, since by an elaborated version of Church's Theorem, there is no recursive set Z separating the formulas with no counter-models from those with finite counter-models.

The foregoing argument applies to just one of the infinitely many formulas that are theorems of **S5** but not of **S4**. Can *all* such formulas be rejected? Clearly, if they can, some general argument, not a case-by-case examination of examples, will be needed to establish that fact. How might such an argument proceed? Well, rejection principles for **S4** have been formulated by Valentin Goranko (1994), and simpler ones have been found by Tomasz Skura (1995), who works with finite algebraic models of the kind alluded to earlier. Skura requires two principles, the first being a slight variant of the rejection principle for **S5** considered earlier.

Unfortunately, Skura's second principle, though simpler than Goranko's principles, is complex enough that it is not very perspicuous, and it is not easy to argue why it should be acceptable for syntactic logical necessity. (Goranko and Skura do not themselves consider such questions.) Fortunately, Skura does *not* claim his rules are the simplest feasible, but on the contrary he explicitly poses it as an open question whether there are any simpler ones. It may be that this open question will have to be settled before one can settle the status of the conjecture that S4 provides the answer to the question of which is the right demonstrability logic.

At present this question, which as I have said goes back to the founders of modern modal logic, remains after most of a century still without a definitive answer.

10

Can truth out?[1]

1 THE DISCOVERY PRINCIPLE

It is rather discouraging that forty years have passed since Frederic Fitch first propounded his paradox of knowability without philosophers having achieved agreement on a solution.[2] As a general rule, when modal phenomena prove puzzling, it is a good idea to look at the corresponding temporal phenomena, and accordingly I propose to examine here not the *knowability* principle that whatever is true *can* be known, but rather the *discovery* principle that whatever is true *will* be known.

As Fitch's modal paradox attacks the knowability principle, so an analogous temporal paradox threatens the discovery principle. The formulation of the paradox is as follows. Start with the minimal tense logic with G and H for "it is always going to be . . ." and "it always has been . . ." as primitive, and F and P for "it sometime will be . . ." and "it once was . . ." defined as ~G~ and ~H~.[3] Add a one-place epistemic operator K for "it is known that," and add as axioms minimal assumptions for this new operator, expressing that anything known is true, and that if a conjunction is known, so are both conjuncts:

(1) $Kp \rightarrow p$
(2) $K(p \,\&\, q) \rightarrow Kp \,\&\, Kq$.

In an attempt to formalize the discovery principle, add one further axiom:

(3) $p \rightarrow FKp$.

[1] First published in Slaerno (2007).
[2] (Fitch 1963). For a summary of recent debates, see B. Brogaard and J. Salerno, "Fitch's Paradox of Knowability," in *Stanford Encyclopedia of Philosophy*: http://plato.stanford.edu/entries/fitch-paradox/ (available only on-line).
[3] See (Burgess 1984a). The various theorems of tense logic cited below can all be found in this source.

The paradox is that one can then derive the following:

(4) $p \rightarrow \mathrm{K}p$.

The derivation of (4) using (3) is, apart from replacing $\Box$ and $\Diamond$ by G and F, the same as Fitch's derivation, which is too well known to bear repeating here.

The operator K is intended to indicate human knowledge, not divine omniscience. The grounds for belief in the discovery principle have indeed traditionally involved a belief in divine omniscience, but it is not this belief alone that supports the principle, but rather this belief plus a further belief that on some future day God will bring it about that whatever is hidden is made manifest (*quidquid latet apparebit*). Obviously that day has not yet come, and the conclusion (4), that everything true is already humanly known, is an absurdity, and so we have a *reductio* of the principle (3).

The "dialethists" and other proponents of radical revisions of classical logic can be counted on to tout their proposed revisions as solutions to this paradox, as they have touted them as the solutions to so many others. But a priori it is overwhelmingly more likely that the problem lies not in the underlying classical logic, but in the least familiar element, the axiom (3), the only axiom in which temporal and epistemic operators interact. And indeed, that *is* where the problem lies. One has to be careful in going back and forth between symbolism and English prose, and Fitch, or rather his hypothetical temporal analogue, was not careful enough.

In tense logic p, q, r, . . . are supposed to stand for tensed sentences, whose truth value may change with time (or if one wants to speak of "propositions," then they must be propositions in a traditional rather than a contemporary sense, propositions that are themselves tensed, and whose truth value may change with time). FA is supposed to be true at a given time if A is true at some later time. What (3) actually expresses thus amounts to this:

(5) If p is true now, then at some later time it will be known that p is true then.

The proposed formalization as (3) has in effect turned the principle that any truth will become known into the principle that any sentence that expresses a truth will come to be known to express a truth. But this last formulation invites the immediate objection that the sentence in question may *cease* to express a truth before the knowledge of the truth it once expressed is acquired.

And so (5) surely does not express what Shakespeare meant in saying "Truth will out." He meant to imply that if Smith murders Jones secretly, so that no one knows, then it will become known that Smith murdered Jones secretly, so that no one knew. He did not mean to imply that if what the form of words "Unknown to all, Smith has murdered Jones" now expresses is true, then there will come a time when what that same form of words then expresses will be known to be true. Thus the temporal analogue of Fitch's argument does not discredit the discovery principle, because the target of that argument is not a correct expression of that principle.

2 INEFFABLE TRUTHS

That one particular objection to a principle fails is no proof of the principle itself, and indeed no proof that it may not be open to simple, straightforward objection along other lines. And in fact the discovery principle is open to two kinds of objection, each of which requires us either to impose a restriction on the principle, or to assume charitably that a restriction on the principle was already intended by its advocates.

As background to a first objection consider the timing of the collision of two ordinary extended material objects. The boundaries of such objects generally are sufficiently ill-defined on a scale of nanometers as to make dating their collision on a scale finer than nanoseconds meaningless. If murders, say, are all the events we want to talk about, we do not need to conceive of "times" as durationless "instants," but may conceive of them as very brief "moments," of no more than, say, a nanosecond's duration. In this case, chronometry – by which I here mean no more than our usual ways of dating events by year, month, day, hour, minute, second, and on to milli- or micro- or nanosecond and beyond if one wishes, all tacitly understood relative to some fixed time zone – supplies a term for every time.

But it may be otherwise if we wish to speak of point-particles and *their* collisions. The worry is that there will be truths that can never be known because they can never be stated. Suppose, for instance, that $\xi = 0.182564793\ldots$ is an irrational number, and that exactly ξ seconds before 12:00 p.m., particle i collided with particle j. Can it ever become known that particles i and j collided at exactly ξ seconds before 12:00 p.m. on June 1, 2003?

According to the discovery principle, all the following will become known:

(1) Particles i and j collided at $182 \pm$ milliseconds before 12:00 p.m. on June 1, 2003.

(2) Particles i and j collided at 182564 ± microseconds before 12:00 p.m. on June 1, 2003.

(3) Particles i and j collided at 182564793 ± nanoseconds before 12:00 p.m. on June 1, 2003.

Here "±" abbreviates "the nearest unit," to the nearest milli- or micro- or nanosecond, as the case may be. (For the sake of argument, set aside any quantum-mechanical doubts about whether the series (1)–(3) really could be continued indefinitely.)

But for it to be knowable that i and j collided at exactly ξ seconds before 12:00 p.m., would it not have to be *sayable* that i and j collided at exactly ξ seconds before 12:00 p.m.? And for this to be sayable, there would have to be some means in language or thought of referring to the irrational number ξ – I mean, of course, some means other than referring to it as the number of seconds before 12:00 p.m. when i and j collided. Mathematics supplies such means for *few* irrational numbers, such as $\sqrt{2}$, π, e, and so forth. Coincidence may supply a few others: the time when i and j collided may be describable also as the time when k and l collide, if the two collisions happen to be simultaneous. But by cardinality considerations we inevitably lack means of reference to *most* irrational numbers.

The discovery principle must be understood to exclude *ineffable* truths. It must be understood as restricted to truths expressible in our language. Such a restriction will be built into any tense-logical formalization of the principle, if the letters p, q, r, . . . are understood as standing for sentences of our language. Such a restriction seems in one sense not too serious, because the principle still tells us that for any question *we have the language to ask*, the true answer will become known.

3 EPHEMERAL TRUTHS

A second objection to the discovery principle is more subtle. Suppose that as I write it is 12:00 p.m., June 1, 2003. Then the following is true:

(1) Now, this moment, it is 12:00 p.m., June 1, 2003.

Obviously (1) itself will never be true in the future. And it seems that no sentence of our language will ever express in the future exactly what (1) expresses now. Thus the truth that (1) now expresses seems to be one that will be unknowable in the future because it is unsayable in the future.

Moreover, the demonstrative "this moment" and the indexical "now" are both pleonastic, what they indicate being already sufficiently indicated by the fact that the verb "is" is in the present tense. Thus what has just been said about (1) is equally true of the following:

(2) It is 12:00 p.m., June 1, 2003.

And indeed, if now, this minute, Smith is murdering Jones, then the following is another example subject to the same difficulty as (1) and (2).

(3) Smith is murdering Jones.

The truth that Smith *is* (now, this moment) murdering Jones seems one that will be unsayable and therefore unknowable in the future, even though it is sayable now and even knowable now.

The discovery principle must be understood to exclude not only *ineffable* truths, which are never expressible in our language, but also *ephemeral* truths, which are expressible for a moment, and then never again. Such a restriction seems in one sense not too serious, because it does not leave us with any question that can always be asked and never be answered. The ephemeral will be equally inexpressible interrogatively as assertorically.

Such a restriction seems not too serious for another reason, because the truths it excludes from human knowledge in the future are excluded even from divine knowledge in eternity, if one follows those theologians who make the latter knowledge timeless. For (1)–(3) are no more true in a timeless eternity than they will be true in the seconds and minutes and hours and days and months and years to come. The old riddle that suggests an exception to the principle that God can see anything I can see is a joke.[4] But the counterexamples (1)–(3) to the principle that God knows anything I can know are not. This point seemed worth digressing to mention, if only because a desire to have a formal apparatus in which such issues could be discussed was an important part of the motivation of the creation of tense logic by Arthur Prior.

4 A REFORMULATION

We have seen that (1.3) – displayed item (3) of §1 – is not the right formalization of the discovery principle. What is? It cannot be claimed

[4] I mean the riddle:

Q. What is it that God never sees, that the king seldom sees, but that you and I see every day?
A. An equal.

This seems less a problem for theologians than for partisans of "substitutional quantification."

that a complete solution to the paradox has been obtained until this question is answered.

One answer suggests itself at once. Now that we have restricted the principle to truths that will remain expressible in our language in the future, it is tempting to formulate the principle as the principle that any sentence that will continue to express a truth in the future will come to be known to express a truth. This goes over into symbols as follows:

(1) $Gp \rightarrow FKp$.

And (1) is, unlike (1.3), immune to Fitch-style paradox, even if one considerably strengthens the background tense logic. For definiteness, let us consider the tense logic, call it L_{linear}, that is appropriate for linearly ordered time without a last time. Then the immunity of (1) from Fitch-style paradox is the content of the following proposition.

Proposition. Let T be L_{linear} plus (1.1), (1.2), and (1). Then (1.4) is not a theorem of T.

Proof. Consider an auxiliary theory T^*, obtained from L by adding a constant π and the following axiom:

(2) $F\pi$.

Then π is not a theorem of T^*. For if we take any of model of L_{linear}, and let π be true at and only at the times later than the present, then (2) will be true at all times, but π will not, being false at all past times and at the present time.

Next assign each formula A of the language of T a translation A^* into the language of T^*, by taking Kp to abbreviate p & π. Thus (1.1), (1.2), (1), and (1.4), respectively, are translated as follows:

(3) $p \,\&\, \pi \rightarrow p$
(4) $(p \,\&\, q) \,\&\, \pi \rightarrow (p \,\&\, \pi) \,\&\, (q \,\&\, \pi)$
(5) $Gp \rightarrow F(p \,\&\, \pi)$
(6) $p \rightarrow p \,\&\, \pi$.

Note that the translation (6) of (1.4) is not a theorem of T^*. For if it were, substituting $\sim\pi$ for p and applying truth-functional logic, π would be a theorem, as we have seen it is not.

To show that (1.4) is not a theorem of T, it will suffice to show that the translation of any theorem of T is a theorem of T^*. And to show this, it will suffice to show that the translations (3)–(5) of the three axioms of T are theorems of T^*. For the first two axioms this is trivial, since (3) and (4) are truth-functional tautologies. For the third axiom, the following is a theorem of L_{linear}:

(7) $Gp \mathbin{\&} Fq \rightarrow F(p \mathbin{\&} q)$.

And (5) follows by truth-functional logic from (2) and (7), to complete the proof.

5 CONSEQUENCES OF THE REFORMULATION

The formalization (4.1) has several corollaries worth noting.

Proposition. Let T be as in §4. Then the following are theorems of T:

(1) $Pp \rightarrow FKPp$
(2) $p \rightarrow FKPp$
(3) $Fp \rightarrow FKPp$
(4) $Gp \rightarrow FKGp$.

Proof. First note that each of the following is either an axiom or a theorem of L_{linear}:

(5) $Pp \rightarrow GPp$
(6) $p \rightarrow GPp$
(7) $Fp \rightarrow FGPp$
(8) $FFp \rightarrow Fp$
(9) $Gp \rightarrow GGp$.

Also, the following is a derived rule of L_{linear}:

(10) If $A \rightarrow B$ is a theorem, then $FA \rightarrow FB$ is a theorem.

(For the cognoscenti, the assumption here is that the rule of temporal generalization, on which (10) depends, continues to apply after the formal language has been enriched by the addition of the epistemic operator K.)

(1), (2), and (4) are immediate from (5), (6), and (9), respectively. As for (3), it can be derived as follows:

(11) $FPp \rightarrow FFKPp$ from (1) by (10)
(12) $FPp \rightarrow FKPp$ from (11) and (8).

6 EXAMPLES

To illustrate these corollaries just derived, if Smith has murdered Jones, or is murdering Jones, or will murder Jones, then according to whichever of

(5.1)–(5.3) is applicable, it will become known that Smith has murdered Jones. Let us write brackets around present-tense verbs to indicate omnitemporality, so that, for instance

(1) Smith [murders] Jones.

is to be understood as meaning

(2) Smith has murdered, is murdering, or will murder Jones.

Then we may say that if Smith [murders] Jones, then it will become known that Smith murdered Jones. And similarly in any other case. Murder cannot be hid – though (4.1) does not go so far as to join the Bard in claiming (unfortunately, erroneously) that murder cannot be hid *long*.

And if the memory of Smith's victim will never cease to be honored, then according to (5.4) this fact will become known – though there is (again, unfortunately) no guarantee it will become known soon enough to comfort the victim's grieving friends and relations. And if the universe will be forever expanding, according to (5.4) this fact, too, will eventually become known – though there is (yet again, unfortunately) no guarantee it will become known soon enough to satisfy the curiosity of present-day cosmologists.

Still, despite its corollaries, (4.1) may look unsatisfactory for the following reason. Consider what the corollary (5.2) tells us about a present truth:

(3) If p is true now, then at some later time it will be known that p was true once.

The "once" here invites the question, "When?" And (5.2) provides no answer.

Or so it may seem. But in a sense (5.2), taken together with chronometry, *does* provide an answer. If (3.2) and (3.3) are true, then the following conjunction is true:

(4) It is 12:00 p.m., June 1, 2003, and Smith is murdering Jones.

Applying (5.2) not to (3.3) alone, but to this conjunction, we obtain

It will become known that it was once 12:00 p.m., June 1, 2003, and Smith was murdering Jones.

or more idiomatically

(5) It will become known that Smith murdered Jones at 12:00 p.m., June 1, 2003.

What more could one want by way of answer to a when-question? Quite generally, an event occurs at a given time, one can conjoin to a sentence p

asserting the event's occurrence a sentence q giving the standard chronometric specification of the *time*, and then apply (5.2), not to p alone, but to the conjunction.

7 "NOW"

Nonetheless, it may seem that the most obvious correction of (1.5) would be the following:

(1) If p is true now, then at some later time it will be known that p was true now.

And (1) seems to tell us more than (4.1) (by way of (6.3)) tells us.

It is known that (1) cannot be expressed using just the temporal operators G and H and F and P. But tense logicians have considered other operators. Most to the point in the present context, they have considered a "now" operator J, so interpreted that even within the scope of a past or future operator Jp still expresses the *present*, not the past or future, truth of p. And with this operator (1) *can* be symbolized, as follows:

(2) $p \rightarrow \mathrm{FKJ}p$.

One may be tempted to think that (2) would do better as a formalization of the discovery principle than does (4.1). But this is a misleading way of putting the issue. For if the operator J is admitted, subject to its usual laws, then (4.1) *implies* (2). For one of the usual laws is precisely

(3) $p \rightarrow \mathrm{GJ}p$

and (2) is immediate from (3) and (4.1). So the temptation here is simply the temptation to add J to the language.[5]

I think the temptation should be resisted for a double reason. My first reason is that introducing the J-operator is unnecessary in order to answer a when-question. For I have just finished arguing that (4.1) does, after all, provide answers to such questions. Against this it may be said that (2) appears to have the advantage of doing so *without* depending on chronometry. But my second reason for avoiding the J-operator is that this apparent advantage comes at the cost of involving us with the problematic notion of a *de re* attitude towards a time.

This truth is perhaps most easily brought to light by switching temporarily from regimentations using tense operators to regimentations using

[5] I owe this observation to Williamson.

explicit quantification over times. So let t, u, v, ... range over times. And let $t < u$ mean that time t is *earlier than* time u, or equivalently, time u is *later than* time t. Let each tensed p be replaced by a one-place $p^*(t)$ for "p [is] the case at time t." Every formula A built up from the letters p, q, r, ... will similarly be replaced by an open formula $A^*(t)$. PA and FA, respectively, will be replaced by

(4a) $\exists u(u < t \,\&\, A^*(u))$ (4b) $\exists u(t < u \,\&\, A^*(u))$.

In a formula $A(t)$ the parameter t may be thought of as standing for that time which is now present. Leaving open how to symbolize the epistemic operator, (5.2) and (2) above go halfway into symbols as follows:

(5) $p(t) \rightarrow \exists u(t < u$ & it is known at time u that $\exists v(v < u \,\&\, p(v)))$
(6) $p(t) \rightarrow \exists u(t < u$ & it is known at time u that $p(t))$.

There is this difference between the two semi-formalizations, that what occurs towards the end of (5) can be understood in a *de dicto* way, thus:

(7) At time u, "p was true once" [is] known to be true.

By contrast, what occurs towards the end of (6) must be understood in a *de re* way, thus:

(8) At time u, "p was true then" [is] known to be true of time t.

The symbol-complex KJp in (2) above may be *pronounced* "it is known that p was true now," but what it really amounts to is more like this:

(9) It is known of t that p was true then, where t is that time which is now present.

8 *DE RE* ATTITUDES

There are (at least) three major difficulties in making sense of the notion of a *de re* knowledge about an object a. Or to put the matter another way, there is only one obvious strategy for making sense of the notion of a *de re* attitude, namely, reduction to a *de dicto* attitude, and there are (at least) three major obstacles to this strategy. The strategy is to understand a subject as knowing of an object a that $F(x)$ holds of it if and only if the subject knows that $F(\alpha)$ where α is a term denoting a. The three obstacles or problems relate to the choice of term α.

A first general problem with *de re* knowledge is that of *anonymity*. There may simply *be* no term α denoting a. This problem has been encountered in the case of times in §2, and given the restriction on the discovery principle imposed there, it may be set aside here.

A second general problem with *de re* knowledge, and one relevant to the question whether J should be admitted, is the problem of *aliases*. The problem is that there may be *two* terms α and β denoting an object *a*, and it may be that the subject knows that $F(\alpha)$ but does not know that $F(\beta)$, or the reverse. The star whose common name is "Aldebaran" has also the official name "Alpha Tauri." It seems that a subject may have been told by different authoritative sources, and hence may know that

(1) Aldebaran is orangish.
(2) Alpha Tauri is the thirteenth brightest star.

and yet, being in ignorance that the two names are names for one and the same heavenly body, may not know that

(3) Alpha Tauri is orangish.
(4) Aldebaran is the thirteenth brightest star.

And this makes it hard to answer the question whether the subject knows of the star itself, independently of how it is named, that *it* is orangish, or the thirteenth brightest. The existence of aliases is a problem insofar as privileging one of them over the other seems *arbitrary*.

The same problem can arise for times. Robinson may know that one rainy day Smith committed murder, and may know that Jones was murdered, and not know that the murder Smith committed was that of Jones. In this case Robinson will know that

(5) At the time when Smith committed murder, it was rainy.

but not that

(6) At the time when Jones was murdered, it was rainy.

And this makes it hard to answer the question whether Robinson knows of the time itself, independently of how it is described, that it was rainy *then*.

Where there exists some *standard* term for each object of a given kind, one can always *stipulate* that a subject is to be credited with *de re* knowledge about the object *a* that $F(x)$ holds of it, if and only if the subject has *de dicto* knowledge that $F(\alpha)$ where α is the *standard* term for *a*. Admittedly, such a stipulation may be more a matter of *giving* a sense to a kind of locution (ascriptions of *de re* knowledge) that previously had none, than of *finding out* what sense this kind of locution had all along. Pretty clearly it would be a case of giving rather than finding if one took as canonical terms for heavenly bodies the official names adopted by international scientific bodies, preferring "Alpha Tauri" over "Aldebaran."

For times, the obvious candidates for standard terms are those provided by chronometry. If one is content with (4.1), there is no need to enter into the problem of *de re* knowledge about times at all, and so no need to fix on any standard terms for times. If one adopts (7.2), reliance on chronometry is the only obvious way to impose a solution on the problem of aliases. But in that case the one advantage (7.2) appeared to have over (4.1), that of *not* depending on chronometry, must be recognized to have been illusory. This consideration argues, I claim, in favor of the J-free formalization (4.1) and against the J-laden formalization (7.2).

A third general problem with *de re* belief is the problem of *demonstratives* (and with them *indexicals*). When the star Alpha Tauri, alias Aldebaran, is visible in the night sky, one can point to it and say "*that* star," and so achieve reference to it. Now it seems someone looking at the star may well know

(5) That star is orangish.

and yet not knowing the name of the star may well *not* know either (1) or (2).

This is, so far, just a special case of the problem of aliases. But demonstratives are especially troublesome because, on the one hand, when available, they seem to provide so direct a way of referring that it is hard to insist that nonetheless it is some *other* way of referring that provides the canonical terms for reduction of *de re* to *de dicto*; but on the other hand, demonstratives themselves are not viable candidates for canonical terms, simply because they are usually *not* available: if we took demonstratives as canonical terms, most objects would suffer from anonymity most of the time. Demonstratives act, so to speak, as spoilers, making any other candidates for the office of canonical term look unworthy, while themselves not being eligible for that office.

But this problem has been encountered in the case of times in §3, and given the restriction on the discovery principle imposed there, it may be set aside here, as the problem of anonymity was set aside. The problem of aliases, I claim, is enough to make the admission of J undesirable.

9 AN IMPERFECT ANALOGY

I have done with the topic of the discovery principle. But what of the knowability principle, and the original, modal version of Fitch's paradox?

Table 10.1

	Temporal	Modal	
	Discovery Principle	Knowability Principle	
	present vs future tense	indicative vs subjunctive mood	
	G, F	$\Box$, $\Diamond$	
	L_{linear}	S5	
(1.3)	$p \rightarrow FKp$	$p \rightarrow \Diamond Kp$	(1)
(4.1)	$Gp \rightarrow FKp$	$\Box p \rightarrow \Diamond Kp$	(2)
(5.2)	$p \rightarrow FKPp$	$p \rightarrow \Diamond K \Diamond p$	(3)
	now	actually	
	J	@	
(7.2)	$p \rightarrow FKJp$	$p \rightarrow \Diamond K@p$	(4)
	times, instants, moments	possibilities, worlds, situations	
	chronometry	???	

I began this essay by recalling that there is a close parallel between temporal and modal. I should now note that while there are many analogies, in connection with Fitch's paradox there is also one glaring disanalogy, that makes the original, modal problem more refractory than its temporal analogue. Perhaps the best way of proceeding would be to begin by simply listing pairs of analogous notions in parallel columns as in Table 10.1.

But returning to what is formally representable, I have recalled in the left margin in the table the numbers of temporal formulas we have met earlier, and assigned in the right margin numbers to the analogous modal formulas. Fitch's (1) is as quickly dismissible as its analogue (1.3).[6] The difficulty comes when one seeks a replacement.

The absence of any obvious analogue for possibilities of standard chronometric specifications for times makes (2) and (3) much less satisfactory than (4.1) and (5.2) – and (4) correspondingly much more tempting than (7.2). But the same absence makes the problem of *de re* knowledge of possible situations connected with (4) at once more critical and more difficult to solve or evade than was the problem of *de re* knowledge of temporal moments connected with (7.2).

I will not enlarge further here, partly because it would be a good exercise for readers to work out the analogy for themselves, but mainly because

[6] For a full exposition of the essentially grammatical fallacy in the paradox, see Rückert (2004). Rückert draws on Wehmeier (forthcoming).

I would be largely repeating points that have been made by Dorothy Edgington in her proposed solution to the paradox, and by Timothy Williamson in his criticisms thereof.[7] A further disanalogy emerges in discussion of Edgington and Williamson that is not formally representable, and is therefore not indicated in Table 10.1. It is just this, that generally speaking the fact that something is only actually true and not necessarily true tends to *matter* less to us than the fact that something is only at the present moment true and not permanently true. Or to put the matter another way, what *will* be true when the world is older matters more to us than what *could have been* true if the world had been otherwise, since we hope to live on into "future worlds" but do not expect to transmigrate into "possible worlds."

So far as the present investigation is concerned, it seems that the analogy between mood and tense takes us only so far, and in the end provides us not with a solution, but only with a better understanding of just what makes the problem difficult.

10 COMBINING TEMPORAL AND MODAL FEATURES

Before giving up, however, perhaps we should try the *combination* of temporal and modal. That is to say, perhaps instead of considering the knowability principle as the principle that anything true could have been known, we should consider it as the principle that anything true could *become* known.

The natural setting for such a principle would be a system like Prior's logic of "historical necessity."[8] In the most elaborate version, which he calls "Ockhamist," Prior uses both tense operators G, H, F, P, subject to the axioms for L_{linear}, and modal operators $\Box$, $\Diamond$, subject to the axioms of S5. But the modal operators are themselves understood in a tensed way, as meaning necessity and possibility *given the course of history up to the present.*

Prior uses special letters *a*, *b*, *c*, ... for sentences with the special property that their truth is independent of the future course of history in addition to the usual letters *p*, *q*, *r*, ... for arbitrary sentences. Not all formulas, but only certain special ones, with the same special property as

[7] See Edgington (1985) and Williamson (1987). For further relevant publications see Brogaard and Salerno (note 2).

[8] See Prior (1967b, chapter VII).

the special letters, may be substituted for those special letters. These include the special letters themselves, any formula beginning with a modal operator $\Box$ or $\Diamond$, and any formula obtainable from formulas of these two kinds using the truth functions and past-tense operators H and P.

A single axiom links the temporal and modal operators:

(1) $a \rightarrow \Box a$.

One can obtain by substitution

(2) $\mathrm{P}a \rightarrow \Box \mathrm{P}a$.

One can*not* derive

(3) $\mathrm{F}a \rightarrow \Box \mathrm{F}a$.

Taking for *a* in (1)–(3) "A sea fight is occurring," in Prior's system one can conclude that if a sea fight is occurring or has occurred, then the occurrence of a sea fight is (historically) necessary; but even if a sea-fight is only *going* to occur, then its occurrence is (historically) contingent, though once it does occur, it will become (historically) necessary.

A version of the knowability principle can be expressed in this context by the formula

(4) $\Box \mathrm{G}a \rightarrow \Diamond \mathrm{FK}a$.

And from (4) one can derive, using various tense logical and modal theorems, the following rough analogues of the corollaries in the proposition of §3:

(5) $(\mathrm{P}a \vee a \vee \mathrm{F}a) \rightarrow \Diamond \mathrm{FKP}a$
(6) $\mathrm{G}a \rightarrow \Diamond \mathrm{FK} \Diamond \mathrm{G}a$.

The details will not be given here, because the system is ultimately unsatisfactory.

Let me explain how. From (4), by way of its corollaries, one can conclude the following, wherein I contract "possibly will" to "may":

(7) If Smith is murdering Jones, then it may become known that Smith has murdered Jones.

(8) If the memory of Smith's victim will always be honored, then it may become known that the memory of Smith's victim may always be honored.

(9) If the universe is always going to be expanding, then it may become known that the universe may be always going to be expanding.

What one can*not* conclude is:

(10) If the memory of Smith's victim will always be honored, then it may become known that the memory of Smith's victim will always be honored.

(11) If the universe is always going to be expanding, then it may become known that the universe *is* always going to be expanding.

So (4) seems too weak.

The strengthening of (4) to

(12) $Ga \rightarrow \Diamond FKGa$

would provide assurance of (10) and (11), but unfortunately (12) is too strong. For it would also provide assurance of the absurd

(13) If Smith murdered Jones but will forever escape detection, then it may become known that Smith murdered Jones but will forever escape detection.

This is Fitch's paradox, adapted to the present context.

II VERIFICATIONISM

In sum, Prior's Ockhamist framework fails to provide a formula that is not too weak and not too strong, but just right. A glance back at the examples above will help us localize the difficulty: it is with truths about the actual future (and more particularly about what will always hold throughout that actual future). There are philosophers, however, who question the very meaningfulness of assertions about the actual future.[9]

And Prior has developed a logic he calls "Peircean" for them. In this logic one has *only* the special letters *a*, *b*, *c*, . . ., and only the formulas built up from them using truth functions, past-tense operators, and four operators amounting to the combinations $\Box G$, $\Diamond G$, $\Box F$, $\Diamond F$. Substitution for the letters *a*, *b*, *c*, . . . is allowed for all formulas so built up. The operators $\Box$, $\Diamond$, G, F do not appear separately, apart from the four combinations just mentioned.

The pertinent feature of this logic in the present context is that it bans as meaningless the examples that caused trouble in the preceding section, and (10.4) seems adequate as an expression of the knowability principle *for all such sentences as are still accepted as meaningful.*

[9] For a recent expression of this view see Belnap and Green (1994).

Banning statements about the actual future is a radical step. Presumably the friends and relations of Jones know that his memory *possibly* will always be honored, and possibly will *not* always be honored. They know $\Diamond Ga$ and $\sim\Box Ga$, where a is

(1) The memory of Smith's victim is honored.

The Peircean, however, rejects as meaningless

(2) The memory of Smith's victim will always be honored.

unless "will" is either strengthened to "necessarily will" or weakened to "possibly will." The Peircean it seems, cannot allow the friends and relations to *hope* that (2) is true, or to *fear* that it is not.[10] Likewise, cosmologists presumably already know that it is *possible* the universe will expand forever, and *possible* that it will not. The Peircean cannot allow them to *wonder* if it in actual fact will.

So Peirceanism is a radical doctrine. But then, so is the knowability principle. The question is, do the two forms of radicalism cohere? If an adherent of the knowability principle were to embrace Peirceanism, would the resulting position have any coherent motivation? Or would embracing Peirceanism be mere ad hoc epicycling, avoiding counter-examples by declaring them meaningless? This is too large, and too non-logical, an issue to go into here, but at least a word may be said about the historical sources of epistemological views like the discovery and knowability principles on the one hand, and of Peirceanism on the other.

Belief in the discovery principle, I said at the outset, has traditionally rested on theological grounds. Belief in the knowability principle has, by contrast, been mainly an expression of a commitment to a certain philosophical theory of meaning, *verificationism*. The radical epistemological view that there are no unknowable truths has usually been a consequence of the even more radical semantical view that understanding a sentence consists in grasping under what conditions it would be known to be true.

Belief in Peirceanism has had several sources. Prior cites late-medieval logicians who have held a similar view on theological grounds, but the more recent proponents of the view seem to base their adherence on grounds that ultimately are verificationist. Thus combining the

[10] This observation is repeated from Burgess (1979d).

knowability principle with Peirceanism could be viewed as combining two manifestations of an underlying verificationism. Of course, there are many varieties of verificationism, and it remains to be seen whether a single variety can cogently motivate both these manifestations simultaneously. A key issue will be the verificationist's attitude towards the reality of the past.

11

Quinus ab omni naevo vindicatus

1 QUINE'S CRITIQUE

1.1 Quine and his critique

Today there appears to be a widespread impression that W. V. Quine's notorious critique of modal logic, based on certain ideas about reference, has been successfully answered. As one writer put it some years ago: "His objections have been dead for a while, even though they have not yet been completely buried."[1] What is supposed to have killed off the critique? Some would cite the development of a new "possible-worlds" model theory for modal logics in the 1960s; others, the development of new "direct" theories of reference for names in the 1970s.

These developments do suggest that Quine's unfriendliness towards any formal logics but the classical, and indifference towards theories of reference for any singular terms but variables, were unfortunate. But in this study I will argue, first, that Quine's more specific criticisms of modal logic have *not* been refuted by either of the developments cited, and further, that there was much that those who did not share Quine's unfortunate attitudes might have learned about modality and about reference by attention to that critique when it first appeared, so that it was a misfortune for philosophical logic and philosophy of language that early reactions to it were as defensive and uncomprehending as they generally were. Finally, I will suggest that while the lessons of Quine's critique have by now in one way or another come to be absorbed by many specialists, they have by no means been fully absorbed by everyone, and in this sense there is still something to be learned from Quine's critique today.

[1] Hintikka (1982, opening paragraph). In context it is clear this is only a description, and not necessarily an endorsement, of a widespread impression.

§3 below will list some lessons from Quine's critique, after §2 has examined the early responses to it. Since I will be arguing that most of these simply missed the point, I should say at the outset that this is easier to see by hindsight than it was from expositions of the critique available at the time, and that the early responses were useful insofar as they provoked new expositions. That there are flaws in Quine's own presentations is conceded even by such sympathetic commentators as Dagfinn Føllesdal and Leonard Linsky, and at least as regards his earliest presentations by Quine himself.[2] To remove flaws is the aim of the present §1, and the aim suggested by my title, which readers familiar with the history of mathematics will recognize as echoing Saccheri's *Euclides ab omni naevo vindicatus* or *Euclid Freed from Every Blemish*. Such readers will also recall that though Saccheri's aim was to defend Euclid, ironically his work is today remembered as a contribution to *non*-Euclidean geometry. While I hope to avoid a similar irony, I do not hesitate to depart from Quine on occasion, and begin with two limitations that I think need more explicit emphasis than they get from Quine.

1.2 Non-trivial de re *modality*

A first restriction is that Quine's critique is limited to predicate as opposed to sentential modal logic, his complaint being that modal predicate logic resulted from mechanically combining the apparatus of classical predicate and modal sentential logic, without thinking through philosophical issues of interpretation.[3] Quine does sometimes suggest that engaging in modal logic would be pointless unless one were eventually going to go beyond the sentential to the predicate level, so that though his critique deals explicitly only with predicate modal logic, it is tantamount to a critique of all modal logic; but the suggestion is not strenuously argued.[4]

The restriction to predicate logic has two aspects. First, the critique is limited to *de re* as opposed to *de dicto* modality, to modalities within the scope of quantifiers as opposed to quantifiers within the scope of

[2] The most important of Quine's presentations is "Reference and modality," in *From a Logical Point of View* (Quine 1953/1961/1980). Citations of this twice-revised work here will be by internal section and paragraph divisions, the same from edition to edition. This work supersedes the earlier Quine (1947a). For commentary see the editor's introduction to Linsky (1971a) and Linsky (1971b). See also Føllesdal (1969) and Føllesdal (1986).

[3] A theme in his reviews (Quine 1946, 1947b).

[4] See the last paragraph of the third section of "Reference and modality," ending: "for if we do not propose to quantify across the necessity operator, the use of that operator ceases to have any clear advantage over merely quoting a sentence and saying that it is analytic."

modalities, to modalities applying to open formulas as in $\exists x \Box Fx$, rather than modalities applying to closed formulas as in $\Box \exists x Fx$. Second, the critique is limited to *non-trivial de re* modality. The first point has been generally understood. Not so the second, which calls for some explanation.

I begin with an analogy. One can contrive systems of sentential modal logic that admit modalities notationally, but that make every modal formula more or less trivially equivalent to a non-modal formula. It suffices to add as a further axiom the following, whose converse is already a theorem in the common systems:

(1) $P \rightarrow \Box P$.

This corresponds to a definition according to which P holds necessarily just in case P holds – a definition that could silence any critic who claimed the notion of necessity to be unclear, but would do so only at the cost of making the introduction of the modal notation pointless.

Analogously, one can contrive systems of predicate modal logic that admit *de re* modalities notationally, but that make every *de re* formula more or less trivially equivalent to a *de dicto* formula. The precise form a *trivialization axiom* would take depends on whether one is considering monadic or polyadic predicate logic, and on whether one is admitting or excluding an existence predicate or an identity predicate or both. In the simplest case it suffices to add as a further axiom the following, whose converse is already a theorem in the common systems:

(2) $\forall x(\Box Fx \rightarrow \Box \forall y Fy)$.

This corresponds to the *trivializing definition* according to which F holds necessarily of a thing just in case it is necessary that F holds of everything – a definition that could silence any critic who claimed the notion of *de re* modality to be more obscure than that of *de dicto* modality, but would do so only at the cost of making the introduction of *de re* notation pointless.

When Quine complains of the difficulty in defining *de re* modality, he is tacitly assuming the trivializing definition above has been rejected; so his critique is tacitly limited to systems that, like all the common ones, do *not* have the trivialization axiom as a theorem. To accept such a system as the correct system, the one whose theorems give all and only the general laws necessarily holding in all instances, is to reject the trivialization axiom as not being such a general law, and hence is to reject the trivializing definition, which would make it one. Note that Quine's objection is thus to the *un*provability of something, namely trivialization, not the *provability* of anything.

1.3 Strict necessity

A second restriction is that Quine's critique is limited to what he calls "strict" necessity, identified with analyticity, as opposed to what may be called "subjunctive" necessity, involved in counterfactuals. For Quine the former belongs to the same circle of ideas as *synonymy* and *definition*, and the latter to the same circle as *similarity* and *disposition*. Quine sometimes explicitly states this limitation; but he also often suggests that his argument generalizes to all intensional operators, or at least that there is an *obstacle* to making sense of quantification into intensional contexts in general (which obstacle is *insurmountable* in the case of quantification into contexts of strict modality in particular).[5] Insofar as I wish to defend it, I take Quine's critique to be limited to strict modality, and his suggestion about generalization to be an attachment to it, not a component of it.

In connection with different senses of necessity there is a feature of the terminology current in the 1940s through 1960s that needs to be explicitly emphasized, lest one fall into anachronistic misreadings: the tendency to use interchangeably with each other, as adjectives modifying the noun "truth," all the expressions in the left-hand column below (and similarly for the right-hand column).[6] Each row merits separate comment.

Necessary	Contingent
Linguistic	Empirical
A priori	A posteriori
Analytic	Synthetic
Logical	Non-logical

Logical truth and analytic truth. Quine distinguished a narrower notion of "logical" truth, roughly truth by virtue of syntactic form alone, from a broader notion of "analytic" truth, roughly truth by virtue of this plus semantic factors such as definition and synonymy. He notoriously thought the latter, broader notion unclear, and so had a *double* objection to the first of the following formulations:

[5] Contrast the opening section of "Reference and modality," on knowledge and belief contexts, with the antepenultimate paragraph of the paper, beginning: "What has been said in these pages relates only to strict modality . . ."

[6] For a contemporary account deploring such tendencies, Kneale and Kneale (1962, pp. 628ff). Such tendencies are exemplified by the usage of *all* the participants in the exchange discussed in §2 below.

(3) It is analytically true that all bachelors are unmarried.
(4) It is logically true that all unmarried men are unmarried.
(3′) "All bachelors are unmarried" is analytically true.
(4′) "All unmarried men are unmarried" is logically true.

One objection was to the common feature of (3) and (3′), involvement with broadly analytic rather than narrowly logical truth; another, to the common feature of (3) and (4), treatment of modality as a connective in the object language applying to sentences, rather than a predicate in the metalanguage applying to quotations. What is important to understand is that in his critique of modal logic Quine presses only his objection to the second feature – a feature presupposed by quantified modal logic, since quantification into quotation contexts is obvious nonsense – waiving his objection to the first for the sake of argument. Others of the period shared neither Quine's worries about the broad, semantic notion, nor his concern to distinguish it from the narrow, syntactic notion, and often wrote "logical" when they meant "analytic."

Analytic truth and a priori truth. Quine's first and foremost target, Rudolf Carnap, and others of the period, took the distinction between analytic and synthetic to be central to epistemology because they took it to coincide with the distinction between a priori and a posteriori. They recognized not a trichotomy of "analytic" and "synthetic a priori" and "a posteriori," but a dichotomy of "analytic" and "a posteriori."

A priori truth and linguistic truth. Quine often complained that others were sloppy about distinguishing use and mention. If one is sloppy, quibbles and confusions can result if, as was commonly done, one uses "linguistic" interchangeably with "analytic" or "a priori" and "empirical" interchangeably with "synthetic" or "a posteriori" respectively. For consider:

(5) Planetoids are asteroids.
(6) Ceres is the largest asteroid.
(5′) In modern English, "planetoids" and "asteroids" refer to the same things.
(6′) In modern English, "Ceres" and "the largest asteroid" refer to the same thing.

As to (5), discovery that planetoids are asteroids requires (for a fully competent speaker of modern English) mere reflection, not scientific investigation. As to (6), discovery that Ceres is the largest asteroid requires natural-scientific investigation of the kind engaged in by astronomers. Discovery that (5′) is the case (understood as about the common language, not just one's personal idiolect) requires social-scientific investigation of the kind engaged in by linguists. Discovery that (6′) is the case requires both kinds of scientific investigation. Since linguistics is an empirical science, using "linguistic" and "empirical" for "analytic" and "aposteriori" can be confusing when dealing with meta-level formulations like (5′) and (6′) rather than object-level formulations like (5) and (6); but such usage was common.

Linguistic truth and necessary truth. Quine distinguished strict and subjunctive modality, but whereas the default assumption today might be that someone who writes "necessary" *sans phrase* intends subjunctive necessity, this was not so for Quine, let alone modal logicians of the period. Originally the primitive notion of modal logic was "implication" $P \multimap Q$, with "necessity" defined as $\neg P \multimap P$; later necessity $\Box P$ was taken as primitive, with implication defined as $\Box\neg(P \wedge \neg Q)$. But even then, the notion of implication of primary interest was strict, so that the notion of necessity of primary interest also had to be, and was often enough explicitly stated to be, strict. It was commonly assumed, if not that *all* necessity is linguistic or semantic or verbal necessity, then at least that the *primary* notion of necessity was that of verbal necessity. In reading the older literature, the default assumption must be that strict necessity is intended when one finds *sans phrase* the word "necessary."

1.4 "Aristotelian essentialism"

Preliminary restrictions having been enumerated, the critique proper begins by indicating what would have to be done to make sense of such notation as $\exists x \Box Fx$. Given that $\exists$ is to be read in what has always been the standard way, as an existential quantifier, and that $\Box$ is to be read in what was at the time the prevailing way, as a strict modality, the following are equivalent:

(7a) $\exists x \Box Fx$ holds
(7b) there is some thing such that $\Box Fx$ holds of it
(7c) there is some thing such that Fx holds necessarily of it
(7d) there is some thing such that Fx holds analytically of it.

The commitment then is to making sense (in a non-trivial way) of the notion of an open formula or open sentence Fx holding analytically of a thing.

Now traditional accounts of analytic truth in philosophy texts provide only an explanation of what it is for a closed sentence to be analytically true, and do not even purport to provide any explanation of a notion of an open sentence being analytically true *of a thing*. (And rigorous analyses of logical truth in logic texts again supply only a definition of what it is for a closed formula to be logically true, and do not even purport to supply any definition of a notion of an open formula being logically true *of a thing*.) The notion of analyticity as it stands simply does not apply *literally* to an open sentence or formula relative *to a thing*, and the most one can hope to do is to extend the traditional notion from *de dicto* to *de re* – or to put the matter the other way round, reduce the notion for *de re* modality to the traditional one for *de dicto* – while remaining faithful to the *spirit* of strict modality. This presumably means remaining attached to a conception of necessity as purely *verbal* necessity, and confined within the circle of ideas containing *definition* and *synonymy* and the like, not bringing in physical notions of *disposition* or *similarity*, let alone Peripatetic or Scholastic metaphysical notions of *matter* and *form* or *potency* and *act* or *essence* and *accident*. Quine expresses pessimism about the prospects for defining *de re* modality subject to this restriction by suggesting that quantified modal logic is committed to "Aristotelian essentialism."

While Quine's own approach is resolutely informal, there is a technical result of Terence Parsons that is illuminating here, even though Parsons' usage of "commitment to essentialism" differs in a potentially confusing way from Quine's. Roughly speaking, Parsons shows that though the common systems are in the sense indicated earlier committed to the failure of trivialization *as a general law*, yet no *specific instance* of such failure is provable in the common systems *even with the addition of any desired consistent set of* de dicto *assumptions*.[7] (On Parsons' usage the result is somewhat confusingly stated as saying that though the common systems are "committed to essentialism" in one, weaker sense, essentially Quine's, they are not "committed to essentialism" in another, stronger sense, Parsons' own.) This being so, any attempt to make sense of *de re* strict modality by reducing it to *de dicto* faces a dilemma.

On the one hand, if one adopts some *general* law permitting passage from *de dicto* to *de re*, one will in effect be adding a new general passage law as an axiom to the common systems. But with any such addition of a new formal axiom one is already rejecting the common systems as incomplete if not as incorrect. Worse, there is a threat that the new axiom will yield trivialization; or worse still, will yield a contradiction. On the other hand, if one allows passage from *de dicto* to *de re* only *selectively*, one will in effect be

[7] For a less rough formulation, see T. Parsons (1969).

adding a new selection principle as an ingredient to the concept of modality. But with any such addition of a new intuitive ingredient there is a danger that one will be making one's conception no longer one of merely *verbal* necessity; or worse, that one will be making it arbitrary and incoherent.

This abstract dilemma is concretely illustrated by Quine's *mathematical cyclist* example, an elaboration of an old example of Mill's, and his *morning star* example, an adaptation of an old example of Frege's. The only obvious approach to reducing the application of modal notions *to a thing* to an application of modal notions *to words*, would be to represent or replace a thing by a word or verbal expression appropriately related to it. In fact, there are two strategies here, the most obvious one being to take the expression to be a *term referring to* the thing, and an only slightly less obvious one being to take the expression to be a *predicate satisfied by* the thing. Hence the need for two examples.

1.5 The mathematical cyclist

One strategy would be to count *Fx* as holding necessarily of a thing just in case *F* is necessarily implied by some predicate(s) *P* satisfied by the thing.

On the one hand, if we are non-selective about the predicates, this leads to contradiction with known or plausible non-modal or *de re* premises, such as the following:

(8a) It is necessarily the case that all mathematicians are rational.
(8b) It is at best contingently the case that all mathematicians are bipeds.
(8c) It is necessarily the case that all cyclists are bipeds.
(8d) It is at best contingently the case that all cyclists are rational.

(These are plausible at least if we take rationality to mean no more than capability for verbal thought, and bipedality to mean no more than having at least two legs, and count mathematicians who have lost limbs as non-bipeds, and count bicycle-riding circus animals as cyclists.) Non-selective application of the strategy to (8a–d) yields:

(9a) Any mathematician is necessarily rational.
(9b) Any mathematician is at best contingently a biped.
(9c) Any cyclist is necessarily a biped.
(9d) Any cyclist is at best contingently rational.

Together (9a–d) contradict the known actual existence of persons who are at once mathematicians and cyclists.

More formally, allowing non-selective application of the strategy amounts to adopting the following as an axiom, which can be seen to collapse modal distinctions all by itself:

(10) $\forall x(Px \rightarrow (\Box Fx \leftrightarrow \Box\forall y(Py \rightarrow Fy)))$.

This is the first horn of the dilemma.

On the other hand, the obvious fall-back would be to allow (10) to apply only selectively, only to certain selected "canonical" predicates. In order for (10), restricted to canonical predicates, to give an adequate definition of *de re* modality, it would suffice for two things to hold. It would suffice to have first that for each thing there is (or can be introduced) some canonical predicate it satisfies; and second that for any two canonical predicates A, B we have:

(11) $\exists x(Ax \wedge Bx) \rightarrow \Box\forall y(Ay \leftrightarrow By)$.

This condition would preclude taking both "x is a mathematician" and "x is a cyclist," or both Plato's "x is a featherless biped" and Aristotle's "x is a rational animal," as canonical. But how is one to select what predicates *are* admitted as canonical? It seems that making a selection, choosing for instance between Plato and Aristotle, would require reviving something like the ancient and medieval notion of "real definitions" as opposed to "nominal definitions"; and this is something it seems impossible to square with regarding the necessity with which we are concerned as simply *verbal* necessity.

1.6 The morning star

The second strategy would be to count Fx as holding necessarily of a thing just in case Ft holds necessarily for some term(s) t referring to the thing.

On the one hand, if we are non-selective about the terms, applying the strategy to all terms equally, then whenever two terms s and t refer to the same thing, Fx holding necessarily of that thing will be equivalent to Fs holding necessarily and equally to Ft holding necessarily, so that Fs holding necessarily and Ft holding necessarily will have to be equivalent to each other. But this result leads to inferences from known or arguably true premises to known or arguably false conclusions, even in the very simple case where Fx is of the form $x = t$, since $t = t$ will in all cases be necessarily true though $s = t$ may in some cases be only contingently true.

For instance, the following are true:

(12) The evening star is the morning star.
(13) Necessarily, the morning star is the morning star.

And the following false:

(14) Necessarily, the evening star is the morning star.

More formally, allowing non-selective application of the strategy amounts to adopting the following as an axiom:

(15) $\forall x(x = t \rightarrow (\Box Fx \leftrightarrow \Box Ft))$.

And this can be seen to collapse modal distinctions (at least if enough apparatus for converting predicates to terms is available). This is the first horn of the dilemma.

On the other hand, the obvious fall-back would be to allow (15) to apply only selectively, to certain selected "canonical" terms. In order for (15), restricted to canonical terms, to give an adequate definition of *de re* modality, two things would be required to hold. It would suffice to have first that for each thing there is (or can be introduced) some canonical term referring to it; and second that for any two canonical terms a, b we have:

(16) $(a = b) \rightarrow \Box (a = b)$.

Now the following is a theorem of the common systems:[8]

(17) $(x = y) \rightarrow \Box (x = y)$.

But (17) involves only variables x, y, . . . , corresponding to pronouns like "he" or "she" in natural language, not constants a, b, . . . or function terms fc, gc, . . . , corresponding to names like "Adam" and "Eve" or descriptions like "the father of Cain" and "the mother of Cain."

[8] It may be worth digressing to mention that Quine's one and only contribution to the formal side of modal logic occurred in connection with this law, though the history does not always emerge clearly from textbook presentations. The earliest derivations of the law took an old-fashioned approach on which identity is a defined second-order notion, and on such an approach the derivation was anything but straightforward. Quine was one of the first to note that on a modern approach with identity a primitive first-order notion, the derivation becomes trivial, and goes through for all systems at least as strong as the minimal normal system **K**. This is alluded to in passing in the penultimate paragraph of the third section of "Reference and modality." For the original presentation see (Barcan 1947). For a modern textbook presentation see Hughes and Creswell (1968, p. 190).

So (17) leaves open what terms should be allowed to be substituted for variables.[9]

What (16) says is that for the fall-back strategy being contemplated to work, we must be able to go beyond (17) to the extent of allowing canonical terms to be substituted for the variables. This condition would preclude taking both "the morning star" and "the evening star" as canonical. But owing to the symmetry involved, it would be entirely arbitrary to select "the morning star" as canonical and reject "the evening star" as apocryphal (or the reverse), and it would seem almost equally arbitrary to reject both and select some other term such as "the second planet." This is the second horn of the dilemma.

And with this observation Quine rests his case, in effect claiming that since the obvious strategies for doing what needs to be done have been tried and found to fail, the burden of proof is now on the other side to show, if they can, just how, in some unobvious way, what needs to be done can be. And with this observation, I too rest my case for the moment.

1.7 Coda

Quine's critique was directed toward the strict kind of modality and toward quantification over ordinary sorts of objects: persons, places, things. Much of his discussion generalizes to other kinds of modal or intensional operators and other sorts of objects, to show that for them, too, the most obvious strategy for making sense of quantifying over such objects into such modal or intensional contexts faces an obstacle. But whether this obstacle can be surmounted, by the most obvious fall-back strategy of identifying an appropriate class of canonical terms or in some other way, needs to be considered case by case. The most important case of a non-strict modality for which a reasonable choice of canonical terms seems to be available (for almost any sort of objects) will be mentioned at the very end of this study. Here I want to mention a case of a special sort of object for which a reasonable choice of canonical terms seems to be available (for almost any kind of intensional operator).

For several writers, beginning with Diana Ackerman, have pointed out that *numerals* suggest themselves as non-arbitrary candidates for canonical

[9] In the original paper where (17) was derived there were no singular terms but variables, and nothing was said about application to natural language. For an idea of the range of options formally available, see the taxonomy in Garson (1984).

terms if one is going to be quantifying only over natural *numbers*. And the numerals are in effect taken as canonical terms in two flourishing enterprises, *intensional mathematics* and *provability logic*, where the modality in question is a version or variant of strict modality.[10]

Still, natural numbers are a *very* special sort of object. Workers in the cited fields have noted the difficulty of finding canonical terms as soon as one goes beyond them even just to other sorts of mathematical objects, such as sets or functions. To avoid difficulties over there simply being *too many* objects to find terms for them all, let us restrict attention to recursively enumerable sets of natural numbers and recursive partial functions on natural numbers, where there is actually a standard way of *indexing* the objects in question by natural numbers or the numerals therefor. Even here there does not seem to be any non-arbitrary way of selecting canonical terms, since there will be many indices for any one set or function, and two indices for the same object will not in general be *provably* indices for the same object.[11]

Whatever successes have been or may be obtained for non-strict modalities and ordinary objects, or for strict modalities and non-ordinary objects, they only make it the more conspicuous how far we are from having any reasonable candidates for canonical terms in the case to which Quine's critique is directed.

2 QUINE'S CRITICS

2.1 *Quine and his critics*

Today when one thinks of model theory for modal logic, or the application of theories of reference to it, one thinks first of Saul Kripke, whose relevant work on the former topic only became widely known after his presentation at a famous 1962 Helsinki conference,[12] and on the latter only after his

[10] See Ackermann (1978). Lectures of Kripke have brought this formerly under-appreciated paper to the attention of a wider audience. See also Shapiro (1985) and especially Boolos (1993, pp. xxxiv and 226).

[11] Workers in the cited fields have in effect suggested that something like indices can serve as canonical terms for more fine-grained intensional analogues of recursive sets and functions. But these too would be very special objects. The best discussion of these matters known to me is in some work not fully published of Leon Horsten.

[12] Whose published proceedings make up a memorable issue of *Acta Philosophica Fennica*, and include not only Kripke (1963) but Hintikka (1963).

celebrated 1970 Princeton lectures.[13] But the impression that somehow an appropriate theory of models or of reference can refute Quine's critique can be traced back a full half-century. For less sophisticated model theories for quantified modal logic go back to some of the first publications on the subject, by Rudolf Carnap, in the 1940s;[14] and the application of less sophisticated theories of reference to modal logic goes back to one of the first reviews of Quine's critical writings, by Arthur Smullyan, again in the 1940s.[15]

For purposes of examining the main lines of response to Quine's critique prior to the new developments in model theory and the theory of reference in the 1960s and 1970s, and Quine's rebuttals to these responses, it is almost sufficient to consider just three documents, together constituting the proceedings of a notorious 1962 Boston colloquium. The main talk, by Quine's most vehement and vociferous opponent, Ruth (Barcan) Marcus, was a compendium of almost all the responses to Quine that had been advanced over the preceding fifteen years, plus one new one. The commentary, by Quine himself, marked an exception to his apparent general policy of not replying directly to critics, and gives his rebuttal to almost all early objections to his critique. An edited transcript of a tape recording of a discussion after the two talks among the two invited speakers and some members of their audience, notably Kripke, was published along with the two papers, and clarifies some points.[16]

2.2 *Potpourri*

A half-dozen early lines of response to the critique may be distinguished. Most appear with differing degrees of explicitness and emphasis in the

[13] Kripke (1972/1980). [14] Carnap (1946, 1947).

[15] Smullyan (1947), with elaboration in Smullyan (1948). Smullyan's priority for his particular response to Quine has been recognized by all competent and responsible commentators. See Linsky (1971b, note 15) and Føllesdal (1969, p. 183).

[16] Thus the items are: (i) the compendium (Marcus 1963a); (ii) the comments (Quine 1963) later retitled "Reply to Professor Marcus"; and (iii) the edited discussion (Marcus, Quine, Kripke *et al.* 1963).They appear together in the official proceedings volume (Wartofsky 1963). The same publisher had printed them in 1962 in *Synthese* in a version that is textually virtually identical down to the placement of page breaks, (i) and (ii) in a belated issue of the volume for 1961, and (iii) in an issue of the volume for 1962. (There have been several later, separate reprintings of the different items, but these incorporate revisions, often substantial.) Two of the present editors of *Synthese*, J. Fetzer and P. Humphreys, have proposed publishing the *unedited, verbatim* transcript of the discussion, with a view to shedding light on some disputed issues of interpretation; but according to their account, one of the participants, Professor Marcus, has objected to circulation of copies of the transcript or the tape.

compendium, and most are rebutted in the commentary thereupon. They all involve essentially the same error, confusing Quine's philosophical complaint with some formal claim. Since – despite the best efforts of Quine himself in his rebuttal and of subsequent commentators – such confusions are still common, it may be in order to review each response and rebuttal briefly.

(A) *The development of possible-worlds semantics shows that there is no problem of interpreting quantified modal logic.* This response is represented in the compendium by the suggestion that disputes about quantified modal logic should be conducted with reference to a "semantic construction," in which connection the now superseded approach of Carnap is expounded (with the now standard, then unpublished, approach of Kripke being alluded to as an alternative in the discussion). Perhaps Quine thought the fallacy in this response obvious, since he makes no explicit response to it in his commentary; but it has proved very influential, albeit perhaps more as an inchoate feeling than as an articulate thought. The fallacy is one of equivocation, confusing "semantics" in the sense of a mathematical theory of models, such as Carnap and Kripke provided, with "semantics" in the sense of a philosophical account of meaning, which is what Quine was demanding, and thus neglecting the dictum that "there is no mathematical substitute for philosophy."[17] A mathematical theory of models could refute a technical claim to the effect that the common systems are formally inconsistent, but without some further gloss it cannot say anything against a philosophical claim that the common systems are intuitively unintelligible. In the case of Carnapian model theory this point perhaps ought to have been obvious from the specifics of the model, which validates some highly dubious theses.[18] In the case of Kripkean model theory the point perhaps ought to be obvious from the generality of the theory, from its ability to accommodate the widest and wildest variety of systems, which surely cannot *all* make good philosophical sense.

[17] These are the closing words of Kripke (1976). The fallacy recurs again and again in other contexts in the literature. See Copeland (1979).

[18] Notably the Barcan or Carnap–Barcan formulas, which give formal expression to F. P. Ramsey's odd idea that whatever possibly exists actually exists, and whatever actually exists necessarily exists. (The "Barcan" label is the more customary, the "Carnap–Barcan" label the more historically accurate according to Cocchiarella (1984), which also explains the connection with Ramsey.) If these formulas are rejected, one must distinguish a thing's having a property *necessarily* (for every possible world it exists there and has the property there) from its having the property *essentially* (for every possible world, *if* it exists there, *then* it has the property there). I have slurred over this distinction so far, and will for the most part continue to do so.

(B) *Quantified modal logic makes reasonable sense if* $\forall$ *and* $\exists$ *are read as something other than ordinary quantifiers, such as Lesniewski-style substitution operators* Π *and* Σ. This is the one substantial novelty in the compendium. One rebuttal, of secondary importance to Quine, is that if one allows oneself to call substitution operators "quantifiers," one can make equally good or poor sense of "quantification" not only into modal but into absolutely any contexts whatsoever, including those of quotation. But quantification into quotation contexts is obvious nonsense – on any *reasonable* understanding of "quantification."[19] Still, the rebuttal of primary importance to Quine is a different and more general one, applying also to the next response.

(C) *Quantified modal logic makes reasonable sense if* $\Box$ *and* $\Diamond$ *are read as something other than strict modalities, such as Prior-style temporal operators* $\mathcal{G}$ *and* $\mathcal{F}$. This response is represented in the compendium by the suggestion, made in passing in the introduction, that modal logic is worth pursuing because of the value of studies of various non-alethic "modalities." The specific example of temporal "modalities" was suggested by Quine in his last remarks in the discussion, his purpose being to bring out his primary point of rebuttal to the previous response, that Lesniewski's devices are just as irrelevant as Prior's devices, given the nature of his complaint. If his complaint had been that there is a formal inconsistency in the common systems, then it would have been cogent to respond by considering those systems as wholly uninterpreted notations, and looking for some reading of their symbolism under which they would come out saying something true or plausible. But the nature of the critique is quite different, the complaint being that the combination $\exists x\Box$ is philosophically unintelligible *when the components* $\exists$ *and* $\Box$ *are interpreted in the usual way*.[20]

(D) *Quantified modal logic is not committed to essentialism because no formula expressing such a commitment* (no instance of the negation of (2)) *is deducible in the common systems, even with the addition of any desired set of consistent* de dicto *axioms*. This response does not explicitly occur as

[19] As shown by examples in the opening section of "Reference and modality." This point seems to be conceded even by some who otherwise take an uncritically positive view of the compendium, as in the review Forbes (1995). The last sections of Kripke (1976) in effect point out that the claim that the ordinary language "there is" in its typical uses is a "substitutional quantifier" devoid of "ontological commitment" is absurd, since "ontological commitment" is *by definition* whatever it is that the ordinary language "there is" in its typical uses conveys.

[20] "What I've been talking about is quantification, in a quantificational sense of quantification, into modal contexts, in a modal sense of modality" (Wartofsky 1963, p. 116).

such in the compendium, and would have been premature, since the results of Parsons which it quotes did not come until a few years later. But it is advanced in a slightly later work of the same author, and has been influential in the literature.[21] It could be construed as merely a generalization of the next response on the list, and Quine's rebuttal to the next response would apply to this one, too. Basically, the response is the result of terminological confusion, since its first clause is only *relevant* if "commitment to essentialism" is understood in Quine's sense, but its second clause is only *true* if "commitment to essentialism" is understood in a different sense partly foreshadowed in the compendium and explicitly introduced as such by Parsons. It has already been noted in the exposition of the critique both that Quine's complaint is not about the *provability* of anything, and that Parsons' results *substantiate* some of Quine's suspicions.

(E) *The mathematical cyclist example does not show there is any problem, because no* de re *conclusions of the kind that figure in the example* (conclusions (9a–d)) *provably follow in the common systems from such* de dicto *premises as figure in the example* (premises (8a–d)). *While the example gives a legitimate counter-instance to the law that figures in it* (law (10)), *that law is not a theorem in the common systems.* This response occurs in a section of the compendium where Quine's criticisms are said to "stem from confusion about what is or is not provable in such systems," and where it is even suggested that Quine believes $\Box(P\rightarrow Q)\rightarrow(P\rightarrow\Box Q)$ to be a theorem of the common systems![22] This response, which accuses Quine of committing a howler of a modal fallacy, is itself a howler, getting the point of Quine's example exactly backwards. The complaint that we cannot deduce examples of non-trivial *de re* modality from plausible examples of non-trivial *de dicto* modality by taking something like (10) as an axiom, because we would get a contradiction, is misunderstood as a formal claim that something like (10) *is* an axiom, and we *do* get a contradiction. Quine's rebuttal in his commentary borders on indignation: "I've never said or,

[21] Marcus (1967). And about the same time we find even the usually acute Linsky (1971a, p. 9) writing: "Terence Parsons bases his search for the essentialist commitments of modal logic on Kripke's semantics, and he comes up (happily) empty-handed . . . He finds modal logic uncontaminated." The continuation of this passage better agrees with Parsons' own account of his work and its bearing on Quine's critique.

[22] See Wartofsky (1963, pp. 90–2). It is just conceivable that this is deliberate exaggeration for effect, a rhetorical flourish rather than a serious exegetical hypothesis. Marcus (1967) cites some other authors who have written in a similar vein about the example.

I'm sure, written that essentialism could be proved in any system of modal logic whatsoever."[23]

(F) *The morning star example does not show there is any problem, because while the law that figures in the example* (law (17)) *is a theorem of the common systems, the example does not give a legitimate counter-instance, as can be seen by applying an appropriate theory of reference.* This response is repeated, with elaboration but without expected acknowledgments – it is described as "familiar," but no specific citation is given – in the compendium. The citation ought to have been to Smullyan.[24] This response again mistakenly takes Quine to be claiming to have a counterexample to a formal theorem of the common systems. (And if Quine *had* claimed that (12) and (14) constitute a counterexample to (17), it would have sufficed to point out that one is not required, just because one recognizes an expression to be a real singular term, to recognize it as legitimately substitutable for variables in all contexts. This point has been noted already in the exposition of the critique, but the response under discussion seems to miss it.) Nonetheless, response (F) is worthy of more extended attention.

2.3 *Smullyanism or neo-Russellianism*

While responses (A)–(E) are entirely skew to Quine's line of argument, response (F) (when fully articulated) makes tangential contact with it, and shows that a minor addition or amendment to the critique as expounded so far is called for. Another reason response (F) calls for more attention than the others is that for a couple of decades it was the conventional wisdom among modal logicians. It was endorsed not only by (in chronological order) Smullyan, Fitch, and Marcus, but also by Arthur Prior and others. It was the topic of two talks at the famous 1962 Helsinki conference and was put forward in major and minor encyclopedias.[25] Yet another reason

[23] And "I did not say that it could ever be deduced in the S-systems or any systems I've ever seen" (Wartofsky 1963, p. 113). Despite these forceful remarks, the understanding of Quine's views has not much improved in the later Marcus (1967).

[24] An earlier paper by the author of the compendium (Marcus 1960) gives a more concise statement of the response in its last paragraph, where a footnote acknowledges the author's teacher Frederic Fitch. The latter, in Fitch (1949) and Fitch (1950), acknowledges Smullyan. (See footnote 4 in the former, footnote 12 in the latter, and the text to which they are attached.)

[25] The major one being Weiss (1967), and the minor one the collection of survey articles (Klibansky 1968). The former contains Prior (1967a) while the latter contains Marcus (1968). The conference talks (Marcus 1963b, Prior 1963) are to be found in the previously cited Helsinki proceedings. Another advocate of closely related ideas has been J. Myhill.

response (F) calls for more attention than the others is that it represents an early attempt to apply a theory of reference distinguishing names from descriptions to the interpretation of modal logic, and understanding why this attempt was unsatisfactory should lead to increased appreciation of more successful later attempts.

The ideas on reference that are involved derive from Russell. The writings of Ramsey, alluded to in passing in the compendium, and of Carnap, with whom the author of the compendium at one time studied, may have served to transmit Russell's influence, though of course Russell himself was still writing on reference in the 1950s, and still living in the 1960s, and should not be considered a remote historical figure like Locke or Mill. But whether his influence on them was direct or indirect, Smullyan's disciples are unmistakably Russell's epigones, even though they seldom directly quote him or cite chapter and verse from his writings.[26]

The Smullyanite response, it will be seen, splits into two parts, one pertaining to descriptions, the other to names. The theory of descriptions presupposed by the Smullyanites is simply the very well-known theory of Russell. The theory of names presupposed is the less well-known theory Russell always took as a foil to his theory of descriptions. This is perhaps best introduced by contrasting it with the theory of Frege, according to which the reference of a name to its bearer is *descriptively mediated*, is accomplished by the name having the same meaning as some description, and the description being uniquely true of the bearer. The theory of Russell is the diametrically opposed one that the reference of a name to its bearer is *absolutely immediate*, in a sense implying that the meaning of a name is simply its bearer, from which it follows that two names having the same bearer have the same meaning. It is taken to follow ("compositionality" being tacitly assumed) that two sentences involving two different names with the same bearer, but otherwise the same, have the same meaning, and hence the same truth value (with one sole exception, usually left tacit, the exception for meta-linguistic contexts, for those sentences, usually involving quotation, where the names are being mentioned as words rather than being used to refer).

This theory is Russell's account of how names *in an ideal sense* would function. While Russell illustrated his theory by examples involving names *in the ordinary sense*, he actually more or less agreed with

[26] Let me not fail to cite chapter and verse myself. For the most relevant pages of the most recently reprinted work, see Russell (1985), pp. 113–15.

Frege about these (so that the Fregean theory is often known as the Frege–Russell theory). Moreover, he held that ordinary, complex things are *not even capable of being given* names in his ideal sense; that names in the ideal sense could be given only to special, simple things (such as sense data). There is an ambiguity running through the writings of all the Smullyanites as to whether they do or do not wish to claim that names in the ordinary sense function as names in the ideal sense. But they do unambiguously wish to claim, contrary to Russell, that whether or not they are already in existence, names in an ideal sense *can at least be introduced* for ordinary things. For this reason, while the Smullyanites may be called "Russellians," it is perhaps better to add the distinguishing prefix "neo-."

So much for the background assumptions of response (F). Its further articulation has several components:

(F0) *Quine's example is ambiguous, since the key terms "the morning star" and "the evening star" might be either mere definite descriptions or genuine proper names.*

(F1a) *If the key phrases are taken to be descriptions, then they are only apparently and not really singular terms, and* (12) *is only apparently and not really a singular identity, so one gets only an apparent and not a real counterexample to* (17).

(F1b) *Moreover, though the foregoing already suffices, it may be added that* (13) *and* (14) *are ambiguous, and it is not unambiguously the case that they are of opposite truth value, the former true and the latter false, as the example claims.*

(F2) *If the key phrases are taken to be names, then* (14) *means the very same thing as, and is every bit as true as,* (13), *contrary to what the example claims.*

To dispose of the issue (F0) of ambiguity, the example may be restated twice:

(12a) Hesperus is Phosphorus.
(13a) Necessarily, Phosphorus is Phosphorus.
(14a) Necessarily, Hesperus is Phosphorus.
(12b) The brightest star of the evening is the brightest star of the morning.
(13b) Necessarily, the brightest star of the morning is the brightest star of the morning.
(14b) Necessarily, the brightest star of the evening is the brightest star of the morning.

2.4 *Quine's rebuttal to neo-Russellianism on descriptions*

The main claim (F1a) of the descriptions side of the Smullyanite response is immediate from Russell's theory, on which (12b) really abbreviates something more complex involving quantifiers:

(12c) There exists a unique brightest star of the evening and
there exists a unique brightest star of the morning, and
whatever is the brightest star of the evening and
whatever is the brightest star of the morning,
the former is the same as the latter.

The subsidiary claim (F1b) is also almost immediate, since on Russell's theory in all but the simplest cases expressions involving descriptions involve ambiguities of "scope," and for instance there is one disambiguation of (14b) that follows by (17) from (12c):

(14c) There exists a unique brightest star of the evening and
there exists a unique brightest star of the morning, and
whatever is the brightest star of the evening and
whatever is the brightest star of the morning,
necessarily the former is the same as the latter.

In rebuttal to all this, the main point is that the example was *not* intended as a counter-instance to (17) or any other theorem of the common systems, but as an illustration of an obstacle to reducing *de re* to *de dicto* modality, so that response (F1a) is wholly irrelevant.

Response (F1b) is partly relevant, however, because it does show that the example needs to be worded more carefully if the Russellian theory of descriptions is assumed. The strategy against which the example was directed was that of defining $\Box Fx$ to hold of a thing if and only if $\Box Ft$ holds where t is a term referring to that thing. But assuming the Russellian theory of descriptions, there is actually more than one strategy here (when t is a description) because $\Box Ft$ is ambiguous between a "narrow" or a "wide" reading.

Also the predicate $\Box Fx$ used in the example, "Necessarily x is the brightest star of the morning" is similarly ambiguous. To eliminate this last ambiguity, take the predicate to be something like "Necessarily, (if x exists then) x is the brightest star of the morning." Then on the narrow-scope reading $\Box Ft$ and $\Box Fs$ boil down to:

(13c) Necessarily, if there exists a unique brightest star of the morning then it is the brightest star of the morning.

(14c) Necessarily, if there exists a unique brightest star of the evening then it is the brightest star of the morning.

So in this case the reduction strategy fails for *the reason originally given*, since (13c) and (14c) are of opposite truth value, the former being true and the latter false. But on the wide-scope reading $\Box Ft$ and $\Box Fs$ boil down instead to:

(13d) There exists a unique brightest star of the morning and necessarily, (if it exists then) it is the brightest star of the morning.

(14d) There exists a unique brightest star of the evening and necessarily, (if it exists then) it is the brightest star of the morning.

In this case the reduction strategy fails for *a more basic reason*, since (13d) and (14d) themselves still involve unreduced *de re* modalities. The claim that the strategy breaks down thus does not have to be retracted, though the explanation *why* it does so needs to be reworded.

Response (F1b) is almost the only significant response to Quine in the early literature not reproduced in the compendium, and for Quine's own statement of a rebuttal to it we need to look beyond his commentary at the colloquium. We find the following formulation, where "non-substitutive position" means a position, such as that of x in $\Box Fx$, where different terms referring to the same thing are not freely intersubstitutable:

> [W]hat answer is there to Smullyan? Notice to begin with that if we are to bring out Russell's distinction of scopes we must make two contrasting applications of Russell's contextual definition of description [as in the (c) versions versus the (d) versions]. But, when the description is in a non-substitutive position, one of the two contrasting applications of the contextual definition [namely, the (d) versions] is going to require quantifying into a non-substitutive position. So the appeal to scopes of descriptions does not justify such quantification, it just begs the question.[27]

2.5 *Neo-Russellianism on names*

Response (F2) is immediate assuming the neo-Russellian theory of names. Indeed, what neo-Russellianism assumes about names is more than enough to guarantee that they would have all the properties required of canonical terms.[28] Thus whereas in rebuttal to (F1) Quine did not have to reject Russell's theory of descriptions, he does have to reject the neo-Russellian theory of names.

Response (F2) is *so* immediate assuming the neo-Russellian theory that it is stated without elaboration by Smullyan and his early disciple Fitch

[27] Reply to Sellars, in Davidson and Hintikka (1969, p. 338). This formulation is the earliest adequate one known to me, the rebuttal even in the 1961 version of "Reference and modality" being inadequate.

[28] As was pointed out in Kripke's last few remarks in the discussion at the colloquium. Quine seems to accept the observation in his last remark. Marcus had apparently ceased to follow by this point.

as if it were supposed to be self-evident.[29] Elaboration is provided by later disciples in the compendium and elsewhere. The elaboration in Prior's talk at the 1962 Helsinki conference is of especial interest because it anticipates in a partial way a significant later contribution to the theory of reference.

Since this has not hitherto been widely noted, I digress to quote the relevant passage:

> It is not necessary, I think, for philosophers to argue very desperately about what is in fact "ordinary" and what is not; but let us say that *a name in Russell's strict sense* is a simple identifier of an object . . .
>
> [T]here is no reason why the same expression, whether it be a single word like "This" or "Tully," or a phrase like "The man who lives next door" or "The man at whom I am pointing," should not be used sometimes as a name in Russell's strict sense and sometimes not. If "The man who lives next door" is being so used, and successfully identifies a subject of discourse, then "The man who lives next door is a heavy smoker" would be true if and only if the subject thus identified *is* a heavy smoker, even if this subject is in a fact a women and doesn't live next door but only works there. And if "Tully," "Cicero," "The Morning Star" and "The Evening Star" are all being so used, then "Tully is Cicero" and "The Morning Star is the Evening Star" both express necessary truths, to the effect that a certain object is identical with itself.[30]

The distinctive part of the passage, not in the founder or other members of the Smullyanite school, is the middle, where it is suggested that even an expression that is *not* a name in the ordinary sense may sometimes *function as* a name. This is a different point from the trivial observation that names often have descriptive *etymologies*, and those familiar with the later literature will recognize how what is said about "the man who lives next door" partially anticipates what was later to be said about "referential" as opposed to "attributive" uses of descriptions.

2.6 *Quine's rebuttal*

The elaboration in Marcus's talk at the same conference, a kind of sequel to the compendium, is of especial interest because it makes more explicit than any other published Smullyanite work the implication that was to be

[29] Fitch (1949) explicitly claims that Quine's contention is "clearly" false if the key expressions are taken to be names.

[30] Prior (1963, pp. 194–5). Prior was from Balliol, and I have heard it asserted – though I cannot confirm it from my own knowledge – that there was a tradition of setting examples of this kind in undergraduate examinations at Oxford in the 1960s.

most emphatically rejected by later work in the theory of reference: the *epistemological* implication that discoveries like (14a) are not "empirical" (at least not in a non-quibbling sense), and are not properly *astronomical* discoveries:

> [T]o discover that we have alternative proper names for the same object we turn to a lexicon, or, in the case of a formal language, to the meaning postulates.... [O]ne doesn't investigate the planets, but the accompanying lexicon.[31]

The same thought had been expressed in slightly different words – "dictionary" for "lexicon," for instance – in the discussion at the colloquium.[32] The picture underlying such remarks had been sketched in the compendium itself:

> For suppose we took an inventory of all the entities countenanced as things by some particular culture through its own language ... And suppose we randomized as many whole numbers as we needed for a one-to-one correspondence, and thereby tagged each thing. This identifying tag is a proper name of the thing.[33]

To talk of an "inventory," and especially to presuppose that we know how many numbers would be "needed for a one-to-one correspondence," is to assume that we are dealing with a known number of unproblematically identifiable items. If it is a matter of applying tags to such items, then of course we should be able to keep a record of when we have assigned multiple tags to a single one of them, though our record would perhaps more colloquially be called a "catalogue" than an "accompanying lexicon" or set of "meaning postulates."

The rebuttal to the Smullyanites on names consists in observing that what is said in the last few quotations is false. Take first Prior. If one defines "names in the strict sense" as expressions with the magical property of presenting their bearers so absolutely immediately as to leave no room for empirical questions of identity, then there never have been in any historically actual language and never can be in any humanly possible language any such things as "names in the strict sense." As Russell himself noted, even "*this* is the same as *this*," where one points to the same object twice, is

[31] Marcus (1963b, p. 132). Note the characteristically Carnapian expression "meaning postulates."

[32] For the published version, too familiar to bear quoting again, see Wartofsky (1963, p. 115). This is one of the parts of the discussion where comparison with the verbatim transcript could be most illuminating. It is a shame that the scholarly public should be denied access to so significant a historical document.

[33] Wartofsky (1963, pp. 83–4). This passage has sometimes been misleadingly cited in the later literature as if it were unambiguously about ordinary names in ordinary language.

not a linguistic and non-empirical truth, if the object in question is complex, and one points to a different component each time.

Take now the compendium and its sequel. Assigning names to heavenly bodies may be like tagging, but it is not like tagging individuals *from among a known number of unproblematically identifiable items*, since we always have unresolved questions before us about the identity of asteroids or comets, as Frege long ago noted. And to resolve such questions one must investigate not some "accompanying lexicon" or "meaning postulates," but the planet(oid)s themselves.

In brief, the following have the same status as (6) and (6′) respectively, and not as (5) and (5′):

(12a) Hesperus is Phosphorus.
(12a′) In modern English, "Hesperus" and "Phosphorus" refer to the same thing.

Quine's own formulation of this rebuttal is almost too well known to bear quotation. But while what Quine *means* is what I have just said, what Quine *says* may be open to quibbles, since taken with pedantic literalness it would seem to be about (12a′) rather than (12a):

> We may tag the planet Venus, some fine evening, with the proper name "Hesperus." We may tag the same planet again, some day before sunrise, with the proper name "Phosphorus." When at last we discover that we have tagged the same planet twice, our discovery is empirical. And not because the proper names were descriptions.[34]

3 QUINE'S LESSONS

3.1 *Hints from Quine for the formal logic of modalities*

With the wisdom of hindsight it can be seen that there are several important lessons about modality and reference directly taught or indirectly hinted in Quine's critique. For modal logic, the first lesson from Quine is that strict or (as many have called it) "logical" modality and subjunctive or (as we now call it) "metaphysical" modality are distinct. A further lesson is that quantification into contexts of strict modality is difficult or

[34] Wartofsky (1963, p. 101). Quine surely means that (12a′) is not just a *linguistic* empirical discovery but a properly *astronomical* empirical discovery. By contrast, Marcus in Wartofsky (1963, p. 115), distinguishes "such linguistic" inquiry as leads to discoveries like (12a′) from "properly empirical" methods such as lead to discoveries about orbits.

impossible to make sense of. A yet further lesson is that quantification into contexts of subjunctive modality is virtually indispensable.

This last lesson is not as explicitly or emphatically taught as the other two, and moreover Quine's remarks are flawed by a tendency to conflate subjunctive or "metaphysical" modality with scientific or "physical" modality – as if we could not speak in the subjunctive of counterfactual hypotheses to the effect that the laws of science or physics were violated. But due allowance being made for this flaw, I believe that the work of Quine, supplemented by that of his student Føllesdal, gives a broad hint pointing in the right direction.

Føllesdal's treatment of the topic begins by quoting and stressing the importance of some of Quine's remarks about the question of the meaningfulness of quantification into contexts of subjunctive modality:

> It concerns ... the practical use of language. It concerns, for example, the use of the contrary-to-fact conditional within a quantification ... Upon the contrary-to-fact conditional depends in turn, for instance, this definition of solubility in water: To say that an object is soluble in water is to say that it would dissolve if it were in water. In discussions in physics, naturally, we need quantifications containing the clause "*x* is soluble in water."[35]

Such passages stop just short of saying, what I think is true, that while quantification into contexts of strict modality may be nonsense, quantification into contexts of subjunctive modality is so widespread in scientific theory and commonsense thought that we could not abandon it as nonsensical even if we wanted to.

Putting the lessons cited together, it follows that there is a difference between strict and subjunctive modality as to what expressions should be accepted as meaningful formulas and so a fortiori as to what formulas should be accepted as correct laws. The strictly or "logically" possible, *what it is not self-contradictory to say actually is*, and the subjunctively or "metaphysically" possible, *what could potentially have been*, differ in the *formalism* appropriate to each.

3.2 *A hint from Quine for the theory of reference of names*

The article on modal logic in the minor encyclopedia alluded to earlier devotes a section to objections, of which the very first (3.1) is Quine's morning star example. In the next section the following is said:

[35] The quotation from Quine is from "Reference and modality," antepenultimate paragraph. The work of Føllesdal where it is quoted is Føllesdal (1965). Føllesdal's final footnote suggests that "causal essentialism" is better off than "logical essentialism," and that Quine's own proposal to treat dispositions as inhering structural traits of objects is a form of "causal essentialism."

> Before proceeding to a summary of recent work in modal logic which is directed toward clear solutions to [such] problems ... it is important to realize that the perplexities about interpretation can only be understood in terms of certain presuppositions held by Quine and others which I will call "the received view" (rv).

A bit later one finds the assertion that: "The Russellian theory of descriptions and the distinction between proper names and descriptions is rejected by rv." This is immediately followed by the assertion that the morning star example (3.1) is "resolved on Russellian analysis as was shown by Smullyan ... and others,"[36] and somewhat later by the insistence that "The usefulness of the theory of descriptions and the distinction between descriptions and purely referential names was argued long before it proved applicable to modal logic," so that one cannot simply reject them, as Quine is alleged to do.

Now some of this account is quite correct, since the theory of descriptions and of the distinction between them and names as one finds it in the compendium, for instance, did not originate there, or even with Smullyan, who first applied it to the interpretation and defense of modal logic, but was indeed argued by Russell long before. But some of this account is quite incorrect. It is *not* true that Quine's rebuttal to Smullyan on descriptions requires rejection of Russell's theory of descriptions.[37] And it is not unambiguously true that Quine's rebuttal to Smullyan on names requires rejection of "the distinction between descriptions and proper names." It is true that it requires rejection of *the neo-Russellian conception* of that distinction, but it is *not* true that Quine insists on rejecting *any* distinction between descriptions and proper names. This should be clear from the last half-sentence of the rebuttal quoted earlier: "And not because the proper names were descriptions."

Before Quine, difficulties with the theory that the reference of a name to its bearer is absolutely immediate had been recognized by Føllesdal and Alonzo Church.[38] And before Quine, difficulties with the theory that the

[36] Klibansky (1968, pp. 91ff). This echoes Fitch (1950, p. 553) where it is said that: "Smullyan has shown that there is no real difficulty if the phrase [*sic*] 'the Morning Star' and 'the Evening Star' are regarded either as proper names or as descriptive phrases in Russell's sense." The syntactic ambiguity in this last formulation as to whether "in Russell's sense" is supposed to modify "proper names" as well as "descriptive phrases" matches the ambiguity in the formulation quoted earlier as to whether "Russellian" is supposed to modify "the distinction between proper names and descriptions" as well as "theory of descriptions." The ambiguity is appropriate, since the theory of names in question is *neo*-Russellian.

[37] Though this may not yet have been made clear at the time the encyclopedia article was written, since the formulation of the rebuttal I have quoted dates from two years later.

[38] See Føllesdal (1961), §17, pp. 96ff) and Church (1950). Both address Smullyan and Fitch.

reference of a name to its bearer is descriptively mediated had also been recognized.[39] But before Quine, those who recognized the difficulties with the absolute immediacy theory generally either did not take them to be decisive or took them to be arguments for the descriptive mediation theory, and vice versa. But if the first lesson of Quine's critique for the theory of reference is that the neo-Russellian theory of names is untenable, the last half-sentence of his rebuttal suggests a second lesson, that this first lesson is *not* in and of itself an argument for the Fregean theory. Putting these lessons together, it is *not* to be assumed that there are just two options; there is space for a third alternative.

3.3 Formal differences between logical and metaphysical modality

A few words may be in order about post-Quinine work on "logical" or strict versus "metaphysical" or subjunctive modalities. The *locus classicus* for the distinction is of course "Naming and necessity," but my concern here will be with *formal* differences, which are not what was of primary concern there. Three apparent such differences have emerged.

First, there is the difference at the predicate level. The conventional apparatus allows *de re* modalities, as in $\Box Rxy$, but does not allow application of different modalities to the different places of a many-place relation. The conclusion that, if one is concerned with logical modality, then the conventional apparatus *goes too far* when it allows *de re* modality, has been endorsed on lines not unrelated to Quine's by a number of subsequent contributors to modal logic, a notable recent example being Hartry Field.[40] The complementary conclusion that, if one is concerned with metaphysical modality, then the conventional apparatus of quantified modal logic *does not go far enough*, when it disallows the application of different modalities to the different places of a many-place relation, has also been advanced by a number of modal logicians, a notable recent example being

[39] For work on difficulties with the Fregean theory in the 1950s and early 1960s, see the discussion in Kripke (1972/1980), and Searle (1967). The doctrines in "Naming and necessity" were first presented in seminars in 1963–4, and whereas that work apologizes for being spotty in its coverage of the literature of the succeeding years, it is pretty thorough in its discussion of the relevant literature (work of P. Geach, P. Strawson, P. Ziff, and others) from the immediately preceding years. (Searle discusses work of yet another contributor, Elizabeth Anscombe.)

[40] In Field (1989, chapter 3). Field also cites several expressions of the same or related views from the earlier literature, and such citations could in a sense be carried all the way back to the "principle of predication" (von Wright 1951).

Max Creswell.[41] (What is at issue in the latter connection is that a two-place predicate *Rxy* may correspond to a phrase with two verbs, such as "*x* is richer than *y* is," each of which separately can be left in the indicative or put in a non-indicative mood, as in "*x* would have been richer than *y* is" contrasting with "*x* would have been richer than *y* would have been," so as to allow *cross-comparison* between how what is is, and how what could have been would or might have been.)

Second, there may well be a formal difference already at the sentential level. For logical modality, at least in some of its versions or variants, *iterated* modalities make good sense. I allude here again to work on intensional mathematics and provability logic, where being unprovable is to be distinguished from being *provably* unprovable. For metaphysical modality, it is much less clear that iteration makes sense. In Prior's well-known work on systems combining subjunctive mood operators with past and future tense operators, for instance, iterated modal operators collapse, unless separated by temporal operators: there is no distinction recognized between what is as of today possibly possible and what is as of today possible, though there is a distinction between what as of yesterday it was possible would be possible as of today and what after all is possible as of today. In later work also on the interaction of mood and tense the purely modal part of the logic adopted amounts to S5, which collapses iterated modalities.[42]

Third, there is the difference that while logical possibility does not admit of degrees – a theory cannot be just a little bit inconsistent – metaphysical possibility seems to, with some possibilities being more remote than others. At any rate, this is the thought that underlies theories of counterfactuals since the pioneering work of R. Stalnaker.[43] In particular, miraculous possibilities, involving violations of the laws of physics, are in general more remote than non-miraculous possibilities, a fact that may make the error of earlier writers in associating counterfactuals with *physical* necessity in some respects a less serious one.

Thus there is a fair amount of work that has been – or can be construed as – exploration of the formal differences between the two kinds of modality.

[41] In Creswell (1990). Cresswell also cites several expressions of the same or related views from the earlier literature, and such citations could in a sense be carried all the way back (Lewis 1970). This is the earliest relevant publication known to me, but its author has suggested that there was very early unpublished work on the topic by A. P. Hazen and by D. Kaplan. The parallel phenomenon for tense in place of mood was noted even earlier by P. Geach.

[42] See Prior (1967b, chapter VII), and among later work R. H. Thomason (1984). The purely modal part is also S5 for virtually all the workers there cited, as well as later ones like A. Zanardo.

[43] Stalnaker (1968). This feature becomes even more prominent in later work on the same topic by D. K. Lewis and others.

As apparent formal differences accumulate, the situation comes to look like this: there is one philosophically coherent enterprise of logical modal logic, attempting to treat in the object language what classical logic treats in the metalanguage; there is another philosophically coherent enterprise of metaphysical modal logic, attempting to do for grammatical mood something like what temporal logic does for grammatical tense; there is a mathematically coherent field of non-classical logics dealing with technical questions about both these plus intuitionistic, temporal, and other logics; but there is no coherent field broad enough to include both kinds of "modal logic," but still narrower than non-classical logic as a whole. In this sense, *there is no coherent enterprise of "modal logic"* – a conclusion that may be called Quinesque.

3.4 New alternatives in the theory of reference for names

A few words may also be in order about post-Quinine work on theories of reference for names that reject both the Fregean *descriptive mediation* and the neo-Russellian *absolute immediacy* views. The *locus classicus* for such an alternative is of course again "Naming and necessity." One can perhaps best begin to bring out how the new theory of that work relates to the old theory of Quine's opponents by considering what similarities and differences are emphasized in the only early extended response to the new theory by the one former adherent of the old theory who remained living and active in the field through the 1970s and 1980s and beyond.[44]

First, the one area of real agreement between the new theory and the old is emphasized, that both are "direct" (in the minimal sense of "anti-Fregean") theories; and the new theory is praised for providing additional arguments:

> Kripke's criticism of the "Frege–Russell" view ... is presented ... Among the arguments he musters are that competent speakers communicate about individuals, using their names, without knowing or being able to produce any uniquely identifying conditions short of circular ones ... Unlike descriptions, proper names are indifferent to scope in modal ("metaphysical") contexts ... Contra Frege he points up the absurdity of claiming that counterfactuals force a shift in the reference of a name.

Second, another area of apparent agreement, over the "necessity of identity" (in some sense), is also emphasized, with the new theory again being praised for providing additional arguments:

[44] Unfortunately this comes in the form of a review of a book by a third party, and is subject to the limitations of such a form. The third party is Leonard Linsky; the book is Linsky (1977); the review is Marcus (1978). The three quotations to follow come from pp. 498, 501, and 502–3.

It is one of the achievements of Kripke's account, with its effective use of the theory of descriptions, the theory of proper names, the distinction between metaphysical and epistemological modalities (for example, necessary vs. a priori), that it provides us with a more coherent and satisfactory analysis of statements which appear to assert contingent identities.

Third, the contribution most praised is the provision of a novel account of the mechanism by which a name achieves reference to its bearer:

> Kripke provided us with a "picture" which is far more coherent than what had been available. It preserves the crucial differences between names and descriptions implicit in the theory of descriptions. By distinguishing between fixing the meaning and fixing the reference, between rigid and nonrigid designators, many nagging puzzles find a solution. The causal or chain of communications theory of names (imperfect and rudimentary as it is) provides a plausible genetic account of how ordinary proper names can acquire unmediated referential use.

All this amounts to something approaching an adequate acknowledgment of substantial *additions* by the new theory to the old, but what needs to be understood is that the new theory in fact proposes substantial *amendments* also. The new theory is not "direct" in anywhere near as extreme a sense as the old. On the new theory, which is a "third alternative," the reference of a name to its bearer is neither descriptively mediated nor absolutely immediate, but rather is *historically mediated*, accomplished through a chain of usage leading back from present speakers to the original bestower of the name. Also the new theory does not endorse the "necessity of identity" in anything like so broad a sense as does the old theory, or on anything like the same grounds. On the new theory, "Hesperus is Phosphorus" is only subjunctively or metaphysically necessary – not strictly or logically necessary like "Phosphorus is Phosphorus." And moreover the metaphysical necessity of identity is the conclusion of a separate argument involving considerations peculiar to subjunctive contexts, about *cross-comparison* between actual and counterfactual situations – not an immediate corollary or special case of some general principle of the intersubstitutability of coreferential names in all (except meta-linguistic) contexts.[45]

The gap between the old, neo-Russellian theory and the new, anti-Russellian theory is large enough to have left space for the development

[45] In this connection mention may be made of one serious historical inaccuracy – of a kind extremely common when authors quote themselves from memory decades after the fact – to be found in the book review, where it is said that the compendium maintained "that unlike different but coreferential descriptions, two proper names of the same object were intersubstitutable in modal contexts" (p. 502). In actual fact, in the compendium it is *repeatedly* asserted that two proper names of the same object are intersubstitutable in *all* contexts.

of several even newer fourth and fifth alternatives, semi- or demi-semi- or hemi-demi-semi-Russellian intermediate views, of which the best known is perhaps Nathan Salmon's.[46] These differ from the Kripkean, anti-Russellian theory in that they want to say that in *some* sense "Hesperus is Phosphorus" and "Phosphorus is Phosphorus" have the same "semantic content."[47] They differ from the Smullyanite, neo-Russellian theory in that there is full awareness that in *some* sense assertive utterance of "Hesperus is Phosphorus" can make a difference to the "epistemic state" of the hearer in a way that assertive utterance of "Phosphorus is Phosphorus" cannot. How it could be that utterances expressing the same semantic content have such different potential effects on epistemic states is in a sense the *main* problem addressed by such theories. My concern here is not to offer any evaluation, or even any exposition, of the solutions proposed, but only to point out that they all operate in the space between Fregeanism and neo-Russellianism – and therefore in a space of whose existence Quine was one of the first to hint.

3.5 Have the lessons been learned?

It would be absurd to claim that Quine anticipated all the many important developments in modal logic or the theory of reference to which I have been alluding. But it is not absurd to suggest that some of them might have been arrived at sooner if the reaction to Quine's critique had been more attentive.

Is the matter of more than antiquarian interest today? Well, certainly there are many workers in philosophical logic and philosophy of language (only a few of whom I have had occasion to mention) who have long since fully absorbed every lesson there was to be learned from Quine. And yet, scanning the literature, it seems to me that specialists in the relevant areas do not always clearly express these lessons in their writings, and that (surely partly in consequence) many non-specialists interested in *applying* theories of modality or reference to other areas have not yet fully learned these lessons.

Take modal logic first. It is said that when Cauchy lectured on the distinction between convergent and divergent series at the Académie des Sciences, Laplace rushed home to check the series in his *Mécanique Céleste.*

46 Salmon (1986). While the early Marcus followed Smullyan, the later Marcus has developed in response to Kripke an idiosyncratic theory that may be described as intermediate in degree of Russellianism between Salmon's and Smullyan's. See Marcus (1990).

47 For Kripke's rejection of this view, see the closing paragraphs of the preface to Kripke (1980).

But when Kripke lectured on the distinction between logical and metaphysical modality, modal logicians did not rush home to check which conclusions hold for the one, which conclusions hold for the other, and which result from a fallacious conflation of the two. It is a striking fact that the basic article – an article written by two very eminent authorities – on modal logic in that standard reference work, the multi-volume encyclopedia mistitled a "handbook" of philosophical logic, makes no mention at all of any such distinction and its conceivable relevance to choosing among the plethora of competing modal systems surveyed.[48]

No wonder then that workers from other areas interested in *applying* modal logic seem often not fully informed about formal differences between the two kinds of modality. To cite only the example I know best, consider philosophy of mathematics, and debates over nominalist attempts to provide a modal reinterpretation of applied mathematics, where quantification into modal contexts is unavoidable. Those on the nominalist side have quite often supposed that they could get away with quantifying into contexts of *logical* modality, while those on the anti-nominalist side have quite often supposed that anyone wishing to make use of modality must stick to the traditional formal systems, which *do not allow for cross-comparison*. Both suppositions are in error.[49]

Take the theory of reference now. Here a great many people seem to have difficulty discerning the important differences among distinct anti-Fregean theories. To mention again the example I know best, many nominalists seem to think that the work of Kripke, David Kaplan, Hilary Putnam, and others has established something implying that it is impossible to make reference to mathematical or other abstract, causally inert objects.[50]

Such misunderstandings are encouraged by the common sloppy use by specialists of ambiguous labels like "causal theory of reference"; and even those who carefully avoid "causal theory" in favor of "direct theory" are often sloppy in their usage of the latter, encouraging other confusions. Of late, not only has the confused opinion become quite common that Quine's critique has somehow been answered by the new theory of

48 Bull and Segerberg (1984). Other articles in the same work, some of which I have already cited, *do* recognize the importance of the distinction.

49 It would be out of place to enter into technicalities here. See Burgess and Rosen (1997).

50 In actual fact, on Kripke's theory, for instance, a name can be given to any object that can be described, not excluding mathematical objects. But again see Burgess and Rosen (1997). (The theory of P. Geach probably deserves and the theory of M. Devitt certainly deserves the label "causal," and does have nominalistic implications.)

names (the one coming from "Naming and necessity"); but so has the even more confused opinion that Quine's critique was already answered by an old theory of names (the one coming from Russell through Smullyan to the compendium); and so too has the most confused opinion of all, that there is no important difference between the old and new theories. Confusion of this kind is found both among those who think of themselves as sympathizers with "the" theory in question,[51] and among those who think of themselves as opponents of "it."[52] The latter cite weaknesses of the old theory as if pointing them out could refute the new theory – a striking example of how confusion over history of philosophy can lead to confusion in philosophy proper.

There is hardly a better way to sort out such confusions than by considering the relations of the old and the new theory to Quine's critique, from which therefore some people still have something to learn. Neither the old theory nor the new provides a refutation of that critique, but the reasons why are radically different in the two cases. The old theory *attempted* to refute that critique, but in doing so it arrived at consequences, notably the one made explicit in the "lexicon" passage quoted earlier, that reduced the theory to absurdity. Quine's rebuttal, pointing out the untenability of these consequences, refuted the old theory. Quine's critique does *not* refute the new theory, but then neither does the new theory refute Quine's critique, nor does it even *attempt* to do so. The new theory *would* refute any incautious claim to the effect that "quantification into any intensional context is meaningless," since it shows that proper names have all the properties required of canonical terms for contexts of *subjunctive* modality. But Quine's critique was addressed to *strict* modality, and as for that, the main creator of the new theory of names has said as I do:[53] "Quine is right."

[51] For comparatively moderate instance see the review Lavine (1995).

[52] For an extreme instance see Hintikka and Sandu (1995). This work acknowledges no important differences among: (i) the neo-Russellian theory of Smullyan as expounded by the early Marcus (which incidentally is erroneously attributed to Marcus as something original, ignoring the real authors Smullyan and Russell); (ii) theories adopted in reaction to Kripke by the later Marcus; and (iii) the theory of Kripke.

[53] In context, what is said to be right is specifically the rebuttal to Smullyanism on names quoted earlier. See Kripke (1972, p. 305).

12

Translating names

I MILLIANISM AND ANTI-MILLIANISM

Mill taught that the signification of a word has in general two components, denotation and connotation, but that in the special case of a proper name there is no connotation, and the signification of the word is just its denotation. According as "meaning" is aligned with "connotation" or with "signification," this doctrine comes out as "A proper name has no meaning" or as "The meaning of a proper name is just its denotation." Today "Millianism" is most often used as a label for the latter version:

(1) The meaning of a name is its denotation.

An immediate consequence of (1) is the following:

(2) Two names with the same denotation have the same meaning.

An immediate objection to (2) is that different names for the same item may be distinguished in level (formal, familiar). Such features are very important for usage. (Imagine what a diplomatic *contretemps* would result if President Chirac were to write President Bush a letter beginning "Yo, Dubya!") And with words that (unlike proper names) appear in dictionaries, such features are commonly noted in their definitions, presumably as part of the meaning of the word. It is therefore, the objector claims, reasonable to take them to be part of the meaning of a name as well.

To this objection a Millian may reply by insisting that level is a feature of usage but not of meaning. This reply illustrates in miniature the fact that in this area there are no conclusive proofs or refutations, but in the end only judgments as to the overall plausibility of proposals as to where to draw the line between semantics and pragmatics. Alternatively, a Millian might concede that (1) and (2) do indeed need to be qualified, while insisting that the objection does not touch what is really important in Millianism, of which the unqualified (1) and (2) are inadequate formulations. This alternative illustrates the fact that it is difficult to find formulations agreeable to all parties of the main theses in dispute. Here I will for brevity let the simple (1) and (2) stand as formulations of Millianism. (Features of level have been mentioned only to illustrate the two facts about the nature of the debate just indicated, and otherwise will play no role in the discussion below.)

I will likewise stick with fairly simple formulations of two other theses, one common to most Millians and anti-Millians alike, the other a premise of one popular form of anti-Millianism. The common thesis is *compositionality*, thus:

(3) Short, simple sentences differing only by substituting one word for another with the same meaning have the same meaning.

Then (2) and (3) together at once give the following:

(4) Short, simple sentences differing only by substituting one name for another with the same denotation have the same meaning.

The anti-Millianism thesis is so-called *transparency*, thus:

(5) When short, simple sentences have the same meaning, a subject's asserting or assenting to one and denying or dissenting from the other is an indication of a deficiency in the subject's linguistic knowledge.

All examples below will involve short, simple sentences (usually three-word sentences of name–copula–adjective form). In particular, none will involve such complications as embedding in "So-and-so believes that . . ." contexts.

The anti-Millian argument I wish to consider produces certain examples, and appeals to pre-theoretic intuitions to support the following claim about them:

(6) There are cases of two short, simple sentences that differ only by substitution of one name for another having the same denotation, where a subject may assert or assent to one sentence and deny or dissent from the other sentence, not for lack of linguistic knowledge, but rather for lack of some other kind of knowledge.

The other knowledge in the best-known examples is biographical or astronomical or geographical or historical (counting etymological knowledge, which is not needed for fluency in the present form of a language, as historical rather than linguistic).

While the examples most often cited pertain to persons (Marcus Tullius Cicero) and planets (Venus), there are also examples about cities. Thus Jones may while doing some tourism have cruised by moonlight past the famous skyline of the former Ottoman capital, and may while negotiating a business deal have paid a rapid visit to an undistinguished commercial district on the outskirts of the same city. If the place was consistently called "Byzantium" by the tour guides and "Istanbul" by the business people, Jones may end up asserting "Byzantium is sublime" and "Istanbul is tacky" while denying "Byzantium is tacky" and "Istanbul is sublime." The anti-Millian's intuition is that this shows Jones to be a geographical ignoramus, but does not show Jones to be a linguistic incompetent like the prize-fighting fan Pete, who applauds whenever someone calls Muhammed Ali "a great boxer" but would take offense if anyone called him "a great pugilist."

This particular case involves binomialism of the first (Hesperus/Phosphorus) kind, where an item has two different names tracing back to two different acts of naming (by Megarian colonists in the seventh century BCE, and by the Atatürk government in the 1930s). But similar examples can arise with binomialism of the second (Cicero/Tully) kind, where two names trace back to the same original act of naming along divergent paths of transmission. A tourist trip to "Peking" may be confined mainly to the Forbidden City, while a business meeting in "Beijing" may be held in a quarter built in the 1950s in high Stalinist style.

2 ANTI-ANTI-MILLIANISM AND TRANSLATION

Now (4) and (5) and (6) cannot all be true, and an accumulation of examples makes (6) hard to deny. The anti-Millian, assuming (5), concludes that the Millian thesis (4) is false. A militant Millian may simply insist that since (4) is true, (5) must be false. But many Millians would, I think, prefer to find some *independent* argument against (5), *not* relying on (4), and one place where they have looked for the materials to construct such an argument has been in Saul Kripke's notorious "A puzzle about belief" (Kripke 1979). That paper contains two examples that, however the author intended them, anti-anti-Millians might appropriate and exploit for purposes of arguing against (5), and it is the anti-anti-Millian appropriation and exploitation of the first of these, the Pierre example, that I wish to consider here.

The example involves the translation of proper names, and so I should acknowledge at the outset the fact that in common parlance one seldom speaks of "translating" proper names at all. In the broad sense used here, whatever expression is used in a translation of a sentence at the place corresponding to the place where a name is used in the sentence being translated may be called a "translation" of that name. But it must be acknowledged that most proper names simply do not have *non-trivial* translations: typically a name is not replaced by *something else* when translating a sentence in which it occurs, but simply taken over for use in the language into which the sentence is being translated, as a so-called *exonym*.

It would be an absurd affectation for a native English speaker, describing to other native English speakers a recent trip to Italy, to speak of having been to "Roma" and "Napoli" and "Firenze." For the famous tourist destinations of Rome and Naples and Florence are among the minority of Italian cities whose names do have non-trivial English translations, or "Anglicizations" as they are more ordinarily called. But if I wish to mention some less famous Italian place, such as Princeton's sister city Pettoranello, there is no alternative to using the Italian name.

More precisely, one uses in speech as close an approximation to the Italian name as one can manage, given the phonetic differences between Italian and English. In writing, one can just use in English the original name if translating from a language written in the Roman alphabet, perhaps modulo some diacritical marks and ligatures. (Thus in the days of old-fashioned manual typewriters, "Gödel" was often written "Goedel," and even today "Gauß" is still written "Gauss.") With languages written in another alphabet one uses what is ordinarily called a "transliteration," and with languages not written alphabetically a "transcription."

The Pierre example involves "London" as the name of the great city in England, which unlike "Liverpool" or "London" as the name of the town in Ontario does have a non-trivial French translation, "Londres." The most direct way to turn the Pierre example into an argument against (5) begins with the following comparatively uncontroversial assumptions:

(7a) "Londres" is the correct French translation of the English "London."
(7b) "est joli[e]" is the correct French translation of the English "is pretty."
(7) "Londres est joli[e]" is the correct French translation of the English "London is pretty."[1]

The next step would be a comparatively uncontroversial inference from (7) to the following:

[1] Should it be "Londres est joli" or "Londres est jolie"? Having consulted with quite a number of informants, French and Québecois, I can report that native Francophones themselves are undecided as to the genders of most city names, "Londres" included.

(8) The meaning of "Londres est joli[e]" (in French) and the meaning of "London is pretty" (in English) are the same.

Kripke then describes the case of one Pierre, of whom he claims the following:

(9) Pierre assents to "Londres est joli[e]" and dissents from "London is pretty."
(10a) There is no deficiency in Pierre's knowledge of French.
(10b) There is no deficiency in Pierre's knowledge of English.
(10) There is no relevant deficiency in Pierre's linguistic knowledge.

Taken together, (8) and (9) and (10) give a counterexample to (5), which may be claimed to neutralize the anti-Millian force of the Byzantium/Istanbul and Peking/Beijing examples, and of the better-known Cicero/Tully and Hesperus/Phosphorus examples.

3 ANTI-ANTI-ANTI-MILLIANISM

I do not wish to question here the truth of (7), or the cogency of the inference thence to (8). Moreover, in the example as Kripke describes it, (9) seems indisputable, too. I think, however, that doubts may be raised about (10), and in order to raise them I wish to question not the truth of but the *grounds* for (7a). So let me begin by mentioning a way to argue for the correctness of translating "London" as "Londres" that would be wholly inappropriate in the present context.

The argument I have in mind takes as its initial premise the one undisputed fact in this area:

(11) "Londres" (in French) and "London" (in English) denote the same place.

It then proceeds to this intermediate step:

(12) "Londres" (in French) and "London" (in English) have the same meaning.

And it then proceeds to the final conclusion (7a). Such an argument would be inappropriate in the present dialectal context, because we are supposed to be looking for an anti-anti-Millian argument *independent* of distinctively Millian assumptions, while it is precisely the distinctively Millian thesis (2) that would be needed to get from (11) to (12).

Questions of dialectic aside, it seems clear that while sameness of denotation is a necessary condition, it is simply not a *sufficient* condition for correctness of translation of place names. If some Francophone counterpart of Jones exclaims, "Que Byzance est belle!" and "Que Stamboul est laid!" he surely has to be translated as saying "How beautiful Byzantium is!"

and "How ugly Istanbul is!" and not "How beautiful Istanbul is!" and "How ugly Byzantium is!" Nor is this just because the person whom we are translating is confused about the identity of the city. "Staline restait à Moscou" and "Djougachvili se cachait" surely have to be translated "Stalin remained in Moscow" and "Djugashvili was hiding," not "Djugashvili remained in Moscow" and "Stalin was hiding," even if the translation is from the writings of Trotsky and is being prepared for a groupuscule of Trotskyites all thoroughly aware that Stalin/Staline and Djugashvili/Djougachvili are one and the same person.

These examples may make it look as if an etymological connection and/or a resultant phonetic or orthographic relationship were the key. But such links are neither necessary nor sufficient for correct translation. Some of the most ancient and famous countries – Egypt, India, Greece – have long been known to most of the outside world by names having nothing to do with their native names. Any one of these cases shows etymological and phonetic and orthographic links are not necessary, as would the case of "Deutschland" aka "Germany" aka "Allemagne." The Greek case shows that they are not sufficient, either. For there actually exists in English a name for the country in question, the name "Hellas," that is etymologically connected and phonetically and orthographically related to the native name; but nonetheless it is the unconnected, unrelated "Greece" that is the correct translation of the native name in all common, prosaic contexts. (Use of "Hellas" and "Hellenic" is appropriate only in the kind of contexts where "Hibernia" and "Hibernian" might be used instead of "Ireland" and "Irish.")

But if not merely denotation, or that plus etymology, what *does* make a given translation of a given name correct or incorrect? A clue is given, I think, by the case of the translation into western European languages – more usually called the "Romanization" – of Chinese proper names. Historically, approximate phonetic transcriptions were used. But since the differences between the principal Chinese "dialects" are so great, and the different European languages are so many, that practice led to chaos, and hence to a demand for a fixed system. Unfortunately not *one* but *two* systems were fixed (both based on Mandarin pronunciation on the Chinese side, but who knows what on the European side), one by European scholars and another by the Chinese government. As a result, most important Chinese geographical places and historical personages can be translated in either of *two* ways into English or French. And often, as with "Hsüan-tsang" aka "Xuánzàng" (the historical personage on whom the character Tripitaka in *A Journey to the West* is based), it is difficult for the average

Anglophone or Francophone reader to trace any link between the Wade–Giles and Pinyin versions. But at least the situation is better than it was when this same personage was also being called by the names "Hiouentang" and "Yuan Chwang" and half a dozen others.

While the case of the Romanization of Chinese, where conventions were consciously and deliberately and explicitly adopted (and where there are two rival sets of conventions), is unusual, I take it to illustrate a more general principle. The general principle is that *what counts as the (or a) correct translation of a name from one language to another is determined by the conventions and customs of the community of bilinguals* and specifically of translators. Needless to say, those conventions and customs are constrained by the requirement that the name and its translation must have the same denotation, and may be influenced, though not determined, by etymological or phonetic or orthographic considerations. In the case of second-, third-, and fourth-hand borrowings, as when, say, an Arabic or Hebrew place-name migrates via Greek and Latin and French to English, suffering distortions at each step, the correct formulation of the general principle would have to be more complicated. (The case where there is no link of any kind, even through a chain of intermediaries, need not be considered, since in this case all names will initially be like "Pettoranello" in lacking non-trivial translations.) But leaving aside details, if anything like this principle is granted, there would seem to follow another important principle, to the effect that *bilingual competence involves something more than just competence in each of the two languages separately, something that can only be acquired by contact with members of the bilingual community*, direct or through their writings.

Now to return to Pierre, the general principle just formulated raises doubts whether (10a) and (10b) are enough to imply (10). Kripke describes Pierre as lacking even the minimal kind of contact with the bilingual community he could acquire by looking into a French–English dictionary. Supposing Pierre keeps a diary in his native French of his experiences in the English city where he finds himself, he will write for the name of that city "London" if he has seen the local name written, or if he has only heard it spoken, something on the order of "Lonnedonne." Either way, a single glance at his diary would suffice to show any bilingual that Pierre lacks the knowledge, which a French–English dictionary could provide him with, that "London" is one of the small minority of English place-names that has a non-trivial French translation, and that this translation is "Londres." I wish to suggest that what this shows is that Pierre is lacking *bilingual* competence, and that this is a lack of a kind of *linguistic* knowledge, making (10) false. Thus at bottom, I suggest, Pierre's state is not after all so very different from that of Ali's linguistically challenged fan Pete.

4 ANTI-MILLIANISM AND ANTI-DESCRIPTIVISM

As we have seen, many examples point to the following principle:

(13) Sameness of denotation of a name in one language with a name in another language is not in general sufficient to make the latter a correct translation of the former.

Many examples also indicate that sameness of denotation of a name in one language and a description in another language are not in general sufficient to make the latter a correct translation of the former. Indeed, something much stronger holds:

(14) A description is never a correct translation of a proper name.

And it should be emphasized that (14) applies to metalinguistic descriptions every bit as much as to any others. It is always wrong to translate "Voltaire" in a French text as "the person called 'Voltaire'" or "the person who called himself 'Voltaire'," and this is so even though "Voltaire" was only a *nom de plume*.

The observations (13) and (14) suggest that Millianism and descriptivism are both alike false. Needless to say, these observations do not *prove* any such conclusions, for the reason I stated at the outset: in any case of apparent counterexample, the militant Millian and the die hard descriptivist can simply suggest that whatever phenomena are cited by their critics pertain to pragmatics rather than semantics. In the present case, the Millian or descriptivist alike could simply claim that sameness of *meaning* is insufficient to make for correct translation. Yet, to repeat, considerations of translation, while not refuting the widespread assumption that Millianism and descriptivism are the only alternatives, do suggest doubts about it. For if one thinks of meaning as that which translation aims to preserve, then a third alternative suggests itself. This third alternative is that different names for the same item have different meanings, even though there is nothing to be said to an English speaker about what "London" (in English) means, except that it means *London* – nothing to be said about what "London" (in English) means that might be informative in the way that it might be informative to tell Pete "'pugilist' means *boxer*," or for that matter to tell Pierre "'London' (en anglais) veut dire *Londres*."

Kripke says that Frege is to be criticized for confusing two senses of "sense." If we ignore Frege's concrete examples, where the senses of names seem to be said to be those of certain associated descriptions, and consider only Frege's abstract formulations, according to which the sense of a name is its mode of presenting its bearer, then the view that different names with the same denotation have different senses certainly seems correct. Presenting a fugitive novelist as "John Doe" and presenting him as "Salmon Rushdie" are certainly, in any ordinary sense of the phrase "way of presenting," two very different ways of presenting him – with perhaps all the difference between life and death between them. The view that different names for the same person or place may have different meanings, even though no name has the same meaning as any description, may thus be restated as the view that different names have different senses in one sense of "senses" (modes of presentation), though they have no senses at all in another sense of "senses" (associated descriptions). Mill's error lies, on this way of putting the matter, not in his claiming that names have no connotation (in the sense of associated descriptions), but rather in his assuming there is nothing more to signification than denotation plus connotation; but this is an error of which descriptivists are guilty as well. In addition to *what* is signified, there is the *way* it is signified, which in the case of names is *not* by connoting some description of it.

However the view is stated, it is one that gains some support from consideration of translation, and one that deserves more attention than it has heretofore been given. It, too, should be taken into account in considering other puzzling cases beyond the kind I have been discussing here. Indeed, the real aim of the present note has been not to solve the puzzle of Pierre, but to call attention to this comparatively neglected view.

5 ENVOI

About the Pierre puzzle there remains one further point to be acknowledged. Suppose Pierre were informed that the French translation of "London" is "Londres." This might well clear up his confusion, but then again it might not. For Pierre might conclude that there are two homonymous names "Londres" in French, one denoting a pretty place he saw in pictures back in France and one the ugly place where he lives now that he has come to England, just as there are two homonymous names "Bretagne," one denoting a peninsula that is part of France, and one denoting a large island across the Channel from France. There is this difference, that in the case of "Bretagne" there are two different names in

English, "Brittany" and "Britain," while in the case of "Londres" there is (so far as Pierre knows) in English just "London." Pierre may be somewhat puzzled why the English have not had the wit to add something to one or both of the names, as the French distinguish their peninsula "Bretagne" from the island "*Grande* Bretagne" or the German town "Aix-la-Chapelle" from the French town "Aix-en-Provence." But perhaps after his ugly experiences living in London he expects no better of the English.

It is clear that if Pierre were to fall into the kind of confusion I have just been describing, then the Pierre example would reduce to a variant of Kripke's *other* example, the Paderewski example. But this example raises a quite different set of issues from the issues about translation with which I have been concerned here, and must be left for another occasion.

13
Relevance: a fallacy?

I INTRODUCTION

Responding to Harvey's theories about the circulation of the blood, Dr. Diafoirus argues (a) that no such theory was taught by Galen, and (b) that Harvey is not licensed to practice medicine in Paris. Plainly there is something wrong with a response of this sort, however effective it may prove to be in swaying an audience. For either or both of (a) and (b) might well be true without Harvey's theory being false. So Diafoirus's argument can serve only to divert discussion from the real question to irrelevant side-issues. The traditional term for such diversionary debating tactics is "fallacy of relevance."

In recent years this tradition has come to be used in a quite untraditional sense among followers of N. D. Belnap, Jr., and the late A. R. Anderson. (All citations of these authors are from their masterwork Anderson and Belnap (1975), and are identified by page number.) According to these self-styled "relevant logicians," it is items (IA) and (IIA) in Table 13.1 that constitute the archetypal "fallacies of relevance." (In the table $\sim$, &, and $\vee$ stand for truth-functional negation, conjunction, and disjunction, respectively.) These forms of argument, say Anderson and Belnap, are "simple inferential mistake[s], such as only a dog would make" (p. 165). The authors can hardly find terms harsh enough for those who accept these schemata: they are called "perverse" (p. 5) and "psychotic" (p. 417).

Needless to say, (IA) and (IIA), which can be traced back at least to Chrysippus, were not traditionally regarded as fallacious. The Anderson–Belnap notion of "relevance," whatever it may amount to, must be something quite different from the traditional notion, which "was central to logic from the time of Aristotle" (p. xxi). And yet the authors declare their so-called "relevant logic" to be a commonsense philosophy, in accord with the intuitions of "naive freshmen" (p. 13) and others who have not been "numbed" (p. 166) by a course in classical logic. Moreover, whereas other

Table 13.1

(I)	*p* or *q*	(II)	not both *p* and *q*
	<u>not *p*</u>		$\underline{p}$
	q		not *q*
(IA)	$p \vee q$	(IIA)	$\sim(p \,\&\, q)$
	$\underline{\sim p}$		$\underline{p}$
	q		$\sim q$
(IB)	$p + q$	(IIB)	$\sim(p \circ q)$
	$\underline{\sim p}$		$\underline{p}$
	q		$\sim q$

dissident logicians (e.g. intuitionists) hold that some forms of argument always accepted and used without question by mathematicians in their proofs are in fact untrustworthy, Anderson and Belnap are at some pains to explain (pp. 17–18 and 261–2) that *their* brand of non-classical logic does *not* conflict with the practice of mathematicians, but only with the classical logician's *account of* that practice.

In view of the fact that everyday arguments and mathematical proofs abound in instances of (I) and (II), one may wonder how Anderson and Belnap could hope to reconcile their objections to (IA) and (IIA) with the claim that their "relevant logic" is compatible with commonsense and accepted mathematical practice. The answer is that the authors believe that ordinary-language argument patterns (I) and (II) should be represented as expressions of the "intensional" schemata (IB) and (IIB), which are relevantistically acceptable, and not of the "extensional" schemata (IA) and (IIA), which relevantists reject.

The compound $p+q$ in (IB) is supposed to be an "intensional disjunction" stronger than the truth-functional $p \vee q$ in that mutual relevance of p and q is required for its truth. This $p+q$ is not entailed by p, nor even by $p \,\&\, q$, since it might be false even though p and q were both true (or even necessary). This happens in the case of irrelevant pairs such as $p =$ "Bach wrote the Coffee Cantata" and $q =$ "The Van Allen belt is doughnut-shaped" (p. 30). Dually, the $p \circ q$ of (IIB) is an "intensional conjunction," better called "cotenability" or "non-preclusion," a compound weaker than the truth-functional $p \,\&\, q$ in that mutual irrelevance of p and q is sufficient for its truth. This $p \circ q$ does not entail p, or even $p \vee q$, since it might be truth even though p and q were both false (or even impossible).

The relevantists' claim that (IB) and (IIB) best represent (I) and (II) admits of two formulations: a stronger and a weaker. The stronger claim would be that the ordinary-language "or" and "and" literally *mean* $+$ and $\circ$ rather than $\vee$ and &. The weaker claim would be that anyone *basing an argument* on the premise that p or q, or that not both p and q, will at least *be in a position to assert* that $p + q$ and $\sim(p \circ q)$ as the case may be. (The latter claim is weaker than the former because even if "or" and "and" *meant* $\vee$ and &, it might still be that arguments of form (IA) and (IIA) could always be avoided in practice because in any instance where one *might wish to argue* from $p \vee q$ or from $\sim(p \,\&\, q)$ the stronger premises $p + q$ and $\sim(p \circ q)$ would *be available.*)

The stronger of these two relevantistic claims seems quite untenable. True, Anderson and Belnap do make a feeble attempt (pp. 176–7) to argue that the ordinary-language "or" usually means $+$ rather than $\vee$. Their argument, however, is scarcely original, amounting to no more than a repetition of the arguments that used to be used by P. F. Strawson and other Oxford philosophers in their diatribes against modern logic. To the serious objections against such Oxonian arguments that have emerged from H. P. Grice's work on conversational implicature the authors of Anderson and Belnap (1975) attempt no reply. In any case, even if the claim that "or" means $+$ is credited with a certain intuitive plausibility, the same cannot be done for the claim that "and" means $\circ$. To be sure, Strawson and others have claimed that the meaning of "and" sometimes diverges from that of & (that in some uses "and" does duty for "and subsequently" or "and as a result"); but these divergences are always in the direction of something *stronger* than &, not something weaker. The relevantists themselves shrink from identifying $\circ$ with "and." While asserting (pp. 344–5) that cotenability is an "analogue" that "in some ways . . . looks like" conjunction, they concede that "it isn't conjunction."

The untenability of the strong claim that everyday and mathematical instances of (I) and (II) are literally *meant* as instances of (IB) and (IIB) might already be thought to do considerable damage to the relevantists' claim to be espousing a commonsense philosophy of logic. My aim in this paper is to cast further doubt on that claim by presenting counterexamples to the weaker claim that everyday and mathematical instances of (I) and (II) can at least be *avoided in favor of* (IB) and (IIB). I will present some examples, taken from everyday life and mathematical practice, of arguments of the forms (I) and (II) which can be neither *read as* nor *replaced by* instances of (IB) and (IIB).

2 EXAMPLES

Background to Example 1

The game of Mystery Cards is played thus: the red and black cards from an ordinary deck are separated. One red and one black "mystery card" are set aside face down, without having been seen by any player. The remaining twenty-five red and twenty-five black cards are combined, shuffled, and dealt out to the players, whose object is to guess the mystery cards. The players take turns questioning each other. The player whose turn it is addresses the player of his choice asking a question of the form, "Is it the such-and-such red card and the thus-and-so black card?" If the player questioned has either or both of the cards named in his hand, he must answer "No"; otherwise he must answer "Maybe." Both question and answer are audible to all players. If a player feels ready to guess the mystery cards, then on his next turn, instead of asking a question he may make a statement, saying "It's the such-and-such red card and the thus-and-so black card!" He then looks at the mystery cards. If his guess is correct, he turns them face up and is declared the winner. If wrong, he puts them back face down, exposes his own hand, and is disqualified from further play.

Admittedly this game is a dull one, but it exhibits in simplified form the principle at work in several more interesting games (e.g. the one marketed under the trade-name CLUE, the importance of which was pointed out to me by D. K. Lewis).

Example 1. Argument

During the course of a game of Mystery Cards, Wyberg hears von Eckes ask Zeemann, "Is it the deuce of hearts and the queen of clubs?" He hears Zeemann reply "No." Later in the game he manages to figure out that it *is* the deuce of hearts. He argues: it isn't both the deuce of hearts and the queen of clubs; but it is the deuce of hearts; so it isn't the queen of clubs. He goes on to use this information to win the game.

Example 1. Analysis

Let $p =$ "The mystery red card is the deuce of hearts," $q =$ "The mystery black card is the queen of clubs." Zeemann's hint is no more and no less than that $\sim(p \,\&\, q)$. Her statement is made on purely truth-functional grounds: she sees the queen in her hand. Her statement is not made on the basis of any "relevance" between p and q: the two mystery cards were chosen entirely independently of each other. Zeemann is justified in denying a truth-functional conjunction, but would not be justified in denying cotenability. Since the premise $\sim(p \circ q)$ is not available to Wyberg, his argument is an instance of (II) that can be neither read as nor replaced by an instance of (IIB).

Had Wyberg been a relevantist, unwilling to make a deductive step not licensed by the Anderson–Belnap systems E and R, he would have been unable to eliminate the queen of clubs from his calculations, and would have lost the game. A relevantist would fare badly in this game and others, and in game-like situations in social life, diplomacy, and other areas – unless, of course, he betrayed in practice the relevantistic principles he espouses in theory.

Background to Example 2

Dr. Zeemann has just been awarded her degree for a dissertation in number theory. Her main result is a proof that every natural number n has either a certain property $A(n)$ or a certain property $B(n)$. As written up in her thesis, the proof is by induction on n, as follows:

Case $n = 0$. We show that $A(0)$. [Here follows a proof.]
Case $n = 1$. We show that $B(1)$. [Here follows a proof.]
Case $n \geq 2$. We assume as induction hypothesis that either $A(n-1)$ and $A(n-2)$, or $A(n-1)$ and $B(n-2)$, or $B(n-1)$ and $A(n-2)$, or else $B(n-1)$ and $B(n-2)$. [Here follows a proof treating each of the four cases separately.]

She remarks that the famous d'Aubel–Hughes Conjecture would imply that $B(0)$, whereas the equally famous conjecture of MacVee would imply that $A(1)$, but reports that she has no light to shed on these old conjectures.

Commentary

Before proceeding, let us note that, following the universal practice of mathematicians, Zeemann has taken her proof that $A(0)$ to dispose of the case $n = 0$ of the general theorem that for all n, either $A(n)$ or $B(n)$. In other words, she argues from the premise $A(0)$ to the conclusion that $A(0)$ or $B(0)$. This is worth mentioning because relevantistically inclined writers have been known to claim that no one ever seriously argues from p to p or q. Indeed, in everyday conversation we are, in R. C. Jeffrey's words, "at a loss to know what the motive could be" for someone to pass from p to the longer and less informative statement that p or q. "Knowing the premise, why not assert *it*, rather than the conclusion?" However, in mathematics we often have good reason to say less than we know: We will assert less than we could about the cases $n = 0$ and $n = 1$ in order to incorporate these cases in a generalization holding for all values of n. Now the inference from $A(0)$ to $A(0)$ or B (0) is only valid if "or" is taken as $\vee$ rather than $+$. Hence Zeemann's theorem must be formalized as $(\forall n)(A(n) \vee B(n))$ and not $(\forall n)(A(n) + B(n))$. This means that any argument of form (I) in which the major premise is supplied by Zeemann's theorem will be an instance of (I) that can be neither read as nor replaced by an instance of (IB). Let us proceed to examples.

Example 2a. Argument

Zeemann applies her work to give bounds to the number of solutions to Tiegh's equation, thus:

> Tiegh himself has shown that the number t of solutions to his equation is ≤ 13. Now a little elementary algebra shows that we cannot have $A(t)$. Hence by our main result, we must have $B(t)$. But no n with $5 \leq n \leq 16$ can satisfy $B(n)$, as is clear from some more elementary algebra. Hence $t \leq 4$.

Example 2a. Analysis

This is a typical mathematical argument of the form (I). The premise $A(t)$ or $B(t)$ must be represented as a truth-functional, not an "intensional" disjunction. The "unknown" t might for all we know be equal to 1, and no "relevant" connection has been established between $A(1)$ and $B(1)$ – indeed, as Zeemann herself reports, she has been unable to establish anything about $B(1)$.

Example 2b. Argument

Professor Wyberg has been working for years on the celebrated conjecture of von Eckes, but has got no further than showing that the conjecture follows from the assumption that $B(1)$, a result he considers not worth publishing. Just recently he has given up work on von Eckes' conjecture in disgust, and has turned to other matters. In particular, he has just refuted an old conjecture of MacVee by proving that $\sim A(1)$. Now he reads an announcement of Zeemann's result. The details of her proof are not available – it takes *years* for theses to come out in print – but he recognizes the significance of her results. In particular, they enable him to prove von Eckes' conjecture at last. He writes a set of notes, "A proof of von Eckes' conjecture," with the following structure: First comes his proof that $\sim A(1)$. Second comes a linking passage:

> And so we see that the MacVee conjecture fails. Now Zeemann has recently announced the result that for all n, either $A(n)$ or $B(n)$. Hence we must have $B(1)$. We now proceed to put this fact to good use.

Third follows the derivation of von Eckes' conjecture from $B(1)$.

Example 2b. Analysis

Since what is established by Zeemann is just $A(1) \vee B(1)$, not $A(1) + B(1)$, we have here another mathematical instance of (I) that can neither be read as nor replaced by an instance of (IB). It is a slightly atypical instance. Had he known the details of Zeemann's work, had he known that she actually proves $B(1)$ outright, Wyberg would surely have just cited this fact that $B(1)$ from her thesis, rather than give the roundabout argument that he did. But this is not to say that the proof of von Eckes' conjecture that Wyberg did give is *erroneous*. One must distinguish inelegance from incorrectness, as even most relevantists allow (p. 279). To sharpen the intuition here, suppose that six months after Wyberg, von Eckes himself notices that his conjecture can be derived from $B(1)$. Suppose further that von Eckes, unlike Wyberg, has access to a photocopy of Zeemann's thesis, and so knows that she has proved $B(1)$. Von Eckes then writes a paper, "Proof of a conjecture in number theory," in which he cites the fact that $B(1)$ from her thesis and then proceeds to derive his old conjecture from $B(1)$ in a manner indistinguishable from that of Wyberg. In this situation, nobody in his right mind would say that von Eckes had produced "the first correct proof" of the conjecture; the honor of priority goes to Wyberg.

One afflicted with relevantistic scruples could not have argued as Wyberg did, but would have had to wait for the publication of Zeemann's work before claiming to have settled von Eckes' conjecture. By that time the less scrupulous Wyberg and the better-placed von Eckes would already be contending for priority. A follower of Anderson and Belnap would not prosper in the world of contemporary mathematics – unless, that is, he sometimes conveniently "forgot" his philosophy of logic.

3 CONCLUSION

No doubt the reader can construct further examples. One might consider, for instance, the case of a person who remembers that once upon a time he was told either that p or that q, but cannot now remember which. Investigating a bit, he quickly establishes that $\sim p$, and so concludes that q. Such examples, I submit, show that as far as negation, conjunction, and disjunction are concerned, "classical" logic (and with it the whole logical tradition from Chrysippus onwards) is far closer to commonsense and accepted mathematical practice than is the "relevant" logic of Anderson and Belnap.

One ploy the relevantist might use in trying to escape from our counter-examples may already have occurred to the reader. What if we take the "relevance" required for the truth of $p + q$ and the falsehood of $p \circ q$ not as something objective and absolute, but as something subjective and relative? We might then say this of the Mystery Cards, for example: in objective fact there is no connection between its being the deuce of hearts and its being the queen of clubs, the red and black cards having been chosen separately. In Zeemann's mind there is no such connection, her statement that it is not both being based solely on her knowledge that it is not the latter. But Zeemann's information establishes such a connection *for Wyberg*, so that he is in a position to assert what she is not, namely $p + q$. Hence his argument can be represented as a case of (IIB).

I doubt such a subjectivization and relativization of "relevance" offers a viable way out to followers of Anderson and Belnap. If (IB) and (IIB) are to cover *all* instances of (I) and (II) in mathematics and everyday argumentation, "relevance" will have to be not just subjectivized but trivialized. *Any* grounds for assertion $p \vee q$ short of the simple knowledge that p, or that q, will have to be taken as sufficient grounds for asserting $p + q$: the statement by a reliable person that she either knows p or knows q though she is not saying which; the knowledge that p holds for $m = 0$ and q holds for $m = 1$, coupled with ignorance as to whether $m = 0$ or 1; the simple recollection that one *once* knew p or knew q though one has now forgotten which. (And paradoxically, the acquisition of *more* information could threaten one's right to assert $p + q$: if one's informant decides to provide more specific information, if the value of m is settled, if one's memory improves, one may suddenly lose the right to assert $p + q$.) Relevantism would reduce to the position that (IA) is valid when and only when one's grounds for asserting $p \vee q$ are something other than the simple knowledge that q. Such a position, however, looks suspiciously like a confusion of the criteria for the *validity* of a form of argument with the criteria for its *utility*, a confusion of logic with epistemology.

Indeed, some writers have been willing to dismiss the whole relevantistic movement as a simple case of confusion between the logical notion of *implication* and the methodological notion of *inference*. The following (unpublished) remarks of G. Harman on this point will bear quoting:

> By reasoning or inference I mean a process by which one changes one's views, adding some things and subtracting others. There is another use of the term "inference" to refer to what I will call "argument", consisting in premises, intermediate steps, and a conclusion. It is sometimes said that each step of an argument should follow from the premises or prior steps in accordance with a "rule of inference". I prefer to say "rule of implication", since the relevant rules do not say how one may modify one's views in various contexts. Nor is there a very direct connection between rules of logical implication and principles of inference. We cannot say, for example, that one may infer anything one sees to be logically implied by one's prior beliefs. Clearly one should not clutter up one's mind with many of the obvious consequences of things one believes.
>
> Furthermore, it may happen that one discovers that one's beliefs are logically inconsistent and therefore logically imply everything. Obviously, one ought not to respond to such a discovery by believing as much as one can. Some philosophers and logicians [the reference is to Anderson and Belnap] have imagined that the remedy here is a new logic in which logical contradictions do not logically imply everything. But this is to miss the point that logic is not directly a theory of reasoning at all.

And indeed *if* "relevance" is taken to be something subjective and relative (according to the proposal discussed above), I do not see how the relevantists could escape Harman's charge that they confuse implication and (useful) inference.

I do not, however, believe that the authors of Anderson and Belnap (1975) understand by "relevance" something subjective. What little they tell us about the nature of "relevance" (e.g. pp. 32–3, where they quote with approval from several sources) strongly suggests that it is a matter of *meaning*. Certainly their commonest charge against classical logic (first raised on p. xxii and repeated ad nauseam) is that it ignores "intension" and meaning. Meaning, however, is something that, generally speaking, will be the same for Wyberg as it is for Zeemann. That relevance is meant to be a semantical, and hence impersonal, notion and not a matter of individual psychology, is further suggested by the relevantists' criticism of T. J. Smiley (p. 217), who is faulted for "epistemologizing" and "psychologizing" the logical notion of entailment. Thus if the authors of Anderson and Belnap (1975) intend by "relevance" something less than objective, they are highly remiss in failing to alert readers to the fact; while if "relevance" is supposed to be impersonal, then the claim that the relevantistic position is (even in a weak sense) compatible with commonsense and accepted mathematical practice succumbs to the counterexamples presented above.

In closing, let me reiterate that I have been concerned here solely with the original Anderson–Belnap account of "relevant" logic, and with their claim that their systems *E*, *R*, etc., are in better agreement with common sense than is classical logic. I have not been concerned with other rationales for developing these systems, nor with the possibility of imposing interpretations on them that were not originally intended by their authors. (It has been suggested, for instance, that some of the formalism created by relevantists might be useful in developing a logic of ambiguity, or of truth-in-fiction.) Workers in category theory, one of the least constructive branches of modern mathematics, have found certain technical uses for intuitionistic logic; but no one imagines that this vindicates Brouwer's philosophy of mathematics. Similarly, the discovery of serendipitous applications of some of the formalism created by Anderson and Belnap would not justify the claim that their logical systems are accurate formalizations of current mathematical practice. Still less could it justify the abusive tone of their remarks about classical logicians.

14

Dummett's case for intuitionism

I TEXTS

Some philosophers approach mathematics saying, "Here is a great and established branch of knowledge, encompassing even now a wonderfully large domain, and promising an unlimited extension in the future. How is mathematics, pure and applied, possible? From its answer to this question the worth of a philosophy may be judged."

Other philosophers approach mathematics in a quite different spirit.[1] They say, "Here is a body, already large and still being extended, of what purports to be knowledge. Is it knowledge, or is it delusion? Only philosophy and theology, from their standpoint prior and superior to that of mathematics and science, are worthy to judge." While this inquisitorial conception of the relation between philosophy and science is less widely held today that it was in Cardinal Bellarmine's time, it continues to have many distinguished advocates.

Prominent among these is Michael Dummett, who has repeatedly advanced arguments for the claim that much of current mathematical theory is delusory and much of current mathematical practice is in need of revision – arguments for the repudiation, within mathematical reasoning, of the canons of classical logic in favor of those of intuitionistic logic. While nearly everything Dummett has written is pertinent in one way or another to his case for intuitionism, there are two texts especially devoted to stating that case: his much anthologized article (Dummett 1973a) on the philosophical basis of intuitionistic logic; and the concluding philosophical chapter of his guidebooks (Dummett 1977) to the elements of intuitionism. The present paper offers a critical examination of these two texts.

[1] For more on the contrast between the two approaches to philosophy of mathematics, see the editorial introduction to Benacerraf and Putnam (1964).

2 DUMMETT'S CASE AGAINST PLATONISM

Dummett has remarked of his case for intuitionism that "it is virtually independent of any considerations relating specifically to the *mathematical* character of the statements under discussion. The argument involve[s] only certain considerations within the theory of meaning of a high level of generality, and could, therefore, just as well have been applied to any statements whatsoever, in whatever area of language" (1973a, p. 226). Hence it is best to begin an examination of his case by considering some of his views on meaning.

Especially important for Dummett are what I will call neutrally *theories of language of the first type*. On a theory of this type, the meaning of a sentence is identified with the conditions for correctness of the sentence (or pedantically: of an assertion made by uttering the sentence) as a representation of reality. A speaker's ability to use the sentence is explained by reference to his grasp of these correctness conditions. Correctness may be conceived of in more than one way, and hence more than one subtype within the first type of theory of language is possible.

Especially important for Dummett is the distinction between those conceptions on which correctness is, and those on which it is not, something always at least in principle potentially recognizable by human beings. The best-known theory operating with a conception of correctness as always recognizable is the *intuitionistic proof-conditional* theory of meaning for the mathematical part of language. (The provability of a mathematical conjecture is recognizable by discovering a proof.) The generalization of this theory to the whole of language would be a *verification-conditional* or *verificationist* theory.

The best-known correctness conception on which correctness is sometimes recognizable, sometimes unrecognizable, is the usual conception of *truth*. (On the usual conception, the truth of a mathematical conjecture need not imply the existence of a proof of the conjecture or any other means of recognizing the conjecture as true.) To avoid ambiguities, I will call truth-as-usually-conceived *verity*. Many distinguished philosophers have advocated *verity-conditional* or *verist* theories of meaning. Dummett calls the specialization of such a theory to the mathematical part of language a *Platonist* theory (though I imagine that he would be hard pressed to locate a passage in the *Republic* or the *Timaeus* where such a theory is taught).

How can graspable correctness conditions be assigned to each of the indefinitely many sentences of a language? The only answer that immediately suggests itself is: inductively. The theories of language of the first type considered by Dummett take the sentences of the language to fall into a hierarchy of degrees of complexity, with the correctness conditions for those of higher degree being determined inductively from the correctness conditions for those of lower degree. For example, such theories include an induction clause indicating how the correctness conditions of a disjunction are determined from those of its disjuncts. On an intuitionist or verificationist theory this clause takes the form:

(1) A proof (or verification) of a disjunction consists in the specification of one of its disjuncts together with a proof (or verification) of that disjunct.

On a Platonist or verist theory this clause takes the form:

(2) A disjunction is true if and only if at least one of its disjuncts is true.

On any theory of language of the first type there is an external standard against which rules of implication are to be judged. A rule is acceptable if and only if it preserves correctness, leading in all instances where the premises are correct to a conclusion that is correct. It is by appeal to such induction clauses as (1) and (2) above that one can seek to demonstrate that certain rules are correctness-preserving or *sound*.

Given the Platonist theory of meaning, the usual soundness proof for classical logic established that all the rules of that logic are acceptable. Given the intuitionist theory of meaning, the usual soundness proof for intuitionistic logic establishes acceptability for all intuitionistic rules, but not for all classical rules: the acceptability of rules depending on the laws of double negation or excluded middle is doubtful when such rules are applied to sentences for which an effective decision procedure is lacking, such as those involving unbounded quantification over an infinite domain. Dummett's strategy is to argue for the repudiation of classical logic in favor of intuitionistic logic by arguing for the repudiation of Platonist (or more generally: verist) theories of meaning in favor of intuitionist (or more generally: verificationist) theories.

On a verist theory, sentences whose correctness need not be recognizable may be said to represent *transcendent* features of reality, while sentences whose correctness must be recognizable may be said to represent *immanent* features of reality. Dummett, like Brouwer, denies that any sentence of any possible language can represent transcendent features of reality. But where Brouwer sees this denial as expressing a limitation on reality, Dummett sees it as expressing a limitation on language.

Dummett's case against verism rests on principles summed up in the slogan that meaning is use. He offers various formulations of these principles, some in terms of *meaning*, others in terms of *understanding*, some telling us what these *consist in*, others through what they are *exhaustively manifested*. The following are typical (1973a, pp. 216, 217):

> The meaning of ... a statement cannot be, or contain as an ingredient, anything which is not manifest in the use made of it, lying solely in the mind of the individual who apprehends that meaning: if two individuals agree completely about the use to be made of the statement, then they agree about its meaning. The reason is that the meaning of a statement consists solely in its rôle as an instrument of communication between individuals ... An individual cannot communicate what he cannot be observed to communicate ...
>
> [T]here must be an observable difference between the behavior or capacities of someone who is said to have ... knowledge [of the meaning of an expression] and someone who is said to lack it. Hence it follows ... that a grasp of the meaning of a ... statement must, in general, consist of a capacity to use that statement in a certain way, or to respond in a certain way to its use by others.

Shared by all such formulations is an association of meaning with public and observable use of language as a vehicle of communication, and a dissociation of meaning from private and hidden use of language as a vehicle of thought.[2] Likewise, any association of meaning or understanding with something in the conscious or unconscious mind, or in the structure or functioning of the brain, is rejected. Though he himself avoids the label, Dummett may be called a *behaviorist* in his approach to meaning, provided this label is understood in a broad enough sense to cover not only the stimulus-response behaviorism of Skinner, but also the logical behaviorism of Ryle.

A thoroughgoing behaviorist will require that any apparatus posited by a semantic theory must be identified or directly correlated with some isolable features of publicly observable verbal behavior. As Dummett formulates it, the behaviorist demand is that there must be a "one-one correspondence between the details" of the apparatus posited by a semantic theory and "observable features of the phenomenon" (1977, p. 377). Dummett describes rejection of this behaviorist demand as one form that the rejection of the principle that meaning is use might take.

The great majority of contemporary linguists reject this behaviorist demand as likely to lead only to sterility and stagnation in semantics. The great majority of contemporary linguists posit in their semantic theories an apparatus neither identified nor directly correlated with any

[2] For more on the contrast between the two sorts of use of language, see the opening paragraphs of Harman (1982).

set of isolable features of publicly observable verbal behavior. The original version of Chomsky's semantic theory, for example, posited an apparatus of *deep structures*. Chomsky and the great majority of contemporary linguists claim that the apparatus posited in their semantic theories is *psychologically real*, represented in ways as yet undiscovered in the mind or brain.[3] But they do not claim the apparatus to be directly represented in behavior.

Thus the principle that meaning is use, on which Dummett bases his case for a revision of current mathematics, itself already amounts to a demand for a revision of current linguistics. For this reason Dummett's arguments for the principle are of interest apart from their role in his case for intuitionism. These arguments have often been criticized by Davidsonians. Here they will be criticized from a viewpoint closer to that of the Chomskians.

Two arguments for the principle that meaning is use are to be found in the texts under examination. They are versions of what have come to be called the *acquisition* and *manifestation* arguments. The first (1973a, p. 217) begins:

> [O]ur proficiency in making the correct use of the statements and expression of the language is all that others have from which to judge whether or not we have acquired a grasp of their meanings. Hence it can only be in the capacity to make a correct use of the statements of the language that a grasp of their meaning, and of those of the symbols and expressions which they contain, can consist.

The rather familiar line of this opening[4] alerts us that the argument is going to turn on how an observer Y, say a teacher, can judge that a speaker X, say a learner, attaches the standard meaning to an expression E. Perhaps before proceeding further it would be well to review schematically the accounts of such judgments offered by behaviorists, on the one hand, and by those anti-behaviorists who associate meaning or understanding with a state of the mind or brain, on the other.

On the behaviorist account, for X to attach the standard meaning to E is for X to be able to use E standardly. On this account, Y's judgment that X

[3] A minority of linguists adopt the position advocated in Soames (1985), regarding such claims of psychological reality as at best premature, but nonetheless insisting on the legitimacy of introducing an apparatus of deep structures or the like in semantic theory, despite behaviorist objections.

[4] Charles Chihara has pointed out in conversation the parallelism between the argument of Dummett just quoted and the notorious argument of Norman Malcolm against the conception of dreams as mental or neural activity taking place at specific times during sleep. Malcolm's argument may be paraphrased: our telling stories when we wake up is all that others have from which to judge whether we have dreamt. Hence it can only be in the disposition to tell stories when we wake up that having dreamt can consist.

attaches the standard meaning to E is a simple inductive inference from the premise that X has been able to use E standardly in all *observed* instances to the conclusion that X will be able to use E standardly in *all* instances. The anti-behaviorist account is much more complex.

The first step towards an anti-behaviorist position is acceptance of the general psychological principle that different people similar in their outward behavior are normally similar also in the inward mental or neural states that causally underlie behavior. The second step is acceptance, as a special linguistic instance, of the hypothesis that there exists a mental or neural state S(E) normally causally underlying the ability to use E standardly. Thus far a behaviorist may or may not go along. Where the behaviorist must refuse to follow is at the anti-behaviorist's third step, the identification of attaching the standard meaning to E with being in the state S(E) rather than directly with being able to use E standardly.

To appreciate the rather subtle distinction here, imagine an abnormal case where a native English speaker X is able to communicate with a native Chinese speaker W only because X has implanted inside his skull a mini-supercomputer programmed to translate back and forth between English and Chinese sentences. There may be no "difference in the behavior or capacities" of X and W that is observable to those of us lacking telepathic powers and X-ray vision. X and W may "agree completely about the use to be made" of various Chinese words and phrases. Yet on account of the absence of "an ingredient . . . lying solely in the mind" or brain, the anti-behaviorist will deny that X attaches the standard meanings, or any meanings at all, to those words and phrases.

The status of such science-fiction examples is in itself a matter of slight importance, but there are more important further differences between behaviorists and anti-behaviorists. One who identifies attaching the standard meaning to E with being the hypothetical state S(E) will presumably be willing to entertain hypotheses about the composition and components of S(E) and to permit a theory of the standard meaning of E to posit an apparatus correlated with these hypothetical components of S(E). But while S(E) itself is normally correlated with the ability to use E standardly, there is no reason to suppose its components to be directly correlated with any "isolable, though interconnected, practical abilities" (Dummett 1977, p. 377). Hence the anti-behaviorist rejection of the requirement that the apparatus posited in a theory of the standard meaning of E must be directly correlated with isolable features of publicly observable verbal behavior, which we have already seen to be the issue dividing Dummettians and Chomskians.

On the anti-behaviorist account, for X to attach the standard meaning to E is for X to be in a mental or neural state S(E) posited to underlie, in normal cases, the ability to use E standardly. On this account, Y's judgment that X attaches the standard meaning to E rests on: (a) the evidence for the presupposition that there exists a mental or neural state underlying, in a normal case, the ability to use E standardly; (b) the evidence that X's case is a normal one; and (c) the evidence that X has been able to use E standardly in all observed instances. The evidence (c) is the only evidence cited in the behaviorist account. The evidence (b) may consist in no more than the absence of evidence that X's case is an abnormal one. The evidence (a) may consist in no more than the evidence for the general psychological principle that different people who are similar in their outward behavior are normally similar also in their inward mental and neural states.

A behaviorist, of course, may question the strength of the evidence for this general psychological principle. Quineans, for example, have claimed that different people identical in their outward behavior may be "like different bushes trimmed to resemble identical elephants." *Dummett's* objections to anti-behaviorism, however, do not take this form, being a priori and philosophical rather than a posteriori and psychological. Returning now to the argument whose opening was quoted above, it continues (1973a, pp. 217–18):

> To suppose that there is an ingredient of meaning which transcends the use that is made of that which carries the meaning is to suppose that someone might have learned . . . [to] behave in every way like someone who understands the language, and yet might not actually understand, or understand it only incorrectly. But to suppose this is to make meaning ineffable, that is, in principle incommunicable. If this is possible, then no one individual ever has a guarantee that he is understood by any other individual; for all he knows, or can ever know, everyone else may attach to his words . . . a meaning quite different from that which he attaches to them. A notion of meaning so private to the individual is one that has become completely irrelevant to mathematics as it is actually practised, namely as a body of theory on which many individuals are corporately engaged, an enquiry within which each can communicate his results to others.

In the earlier parts of this passage, Dummett claims that if there is anything more to X's attaching the standard meaning to E than X's being able to use E standardly, then Y can never know or have a guarantee that X attaches the standard meaning to E. Two comments are called for. First, on *both* the anti-behaviorist *and* the behaviorist accounts, Y's judgment that X attaches the standard meaning to E is an inductive inference. On the behaviorist account, it is an inference from a limited number of observed instances of use to an unlimited number of possible future instances of use. No inductive inference can provide *certain* knowledge or an *indubitable* guarantee.

But if the possibility of skeptical doubt and uncertainty somehow undermines a theory of meaning, it must undermine not just the anti-behaviorist theory, on which understanding is something mental or neural transcending behavior, but also the behaviorist theory, on which understanding is an open-ended behavioral ability or capacity, transcending any finite number of its manifestations. This point has been mentioned in passing by Susan Haack (1974, pp. 107–8) and developed at length by Crispin Wright (1980, pp. 123–8).

Second, it is far from obvious that the impossibility of skepticism-proof guaranteed knowledge in any way undermines a theory of meaning. In the later parts of the passage under examination, Dummett seems to try to draw out damaging consequences from the absence of such guaranteed knowledge. He seems to claim that its absence somehow makes *communication* between mathematicians impossible, and hence makes *mathematics* as an activity involving communication impossible. Surely such a claim would be mistaken. For whether mathematicians X and Y succeed in communicating through their use of expression E surely depends only on whether X and Y *do in actual fact* attach the same meaning to E, and not on whether they possess skepticism-proof guaranteed knowledge that they do so. One hesitates to accuse a distinguished authority on modal logic of arguing from $\Diamond{\sim}p$ to ${\sim}\Diamond p$, but Dummett does almost seem to wish to move from the (epistemic) possibility that X and Y do not succeed in communicating to the (metaphysical) impossibility of X and Y succeeding in communicating.

To avoid fallacies in the "acquisition" argument we must distinguish the claim that a language learner cannot come to attach the standard meaning to an expression from the claim that no one can have guaranteed knowledge that the language learner has attached the standard meaning to the expression. Dummett fails to show that the former follows from the anti-behaviorist approach to meaning; and while the latter may follow, it is not obviously unacceptable. This last point has been noted by Dag Prawitz, who writes (1977, p. 10) of Dummett's argument that:

> One could contest it by arguing that when we learn a language by seeing how its sentences are used, we only get some hints about their meaning. The samples of use with which we are presented never completely determine the meaning but only enable us to form some theories or hypotheses about the meaning. (The fact that we nevertheless agree rather well about meaning could perhaps be explained by

reference to a genetic disposition to see certain kinds of patterns and hence to form certain kinds of theories upon seeing a few examples.) Such a view would entail that we could never be sure that we knew the meaning of a sentence; a new unexpected use of it could show us that we had misunderstood the meaning and would force us to revise our theory. And to some extent this may be a correct picture of our situation.

Dummett's other argument calls for less comment. It runs as follows (1977, p. 217):

Now knowledge of meaning ... is frequently verbalisable knowledge, that is, knowledge which consists in the ability to state the rules in accordance with which the expression or symbol is used ... But to suppose that, in general, a knowledge of meaning consisted in verbalisable knowledge would involve an infinite regress: if a grasp of the meaning of an expression consisted, in general, in the ability to *state* its meaning, then it would be impossible for anyone to learn a language who was not already equipped with a fairly extensive language. Hence that knowledge which, in general, constitutes the understanding of language ... must be implicit knowledge. Implicit knowledge cannot, however, meaningfully be ascribed to someone unless it is possible to say in what the manifestation of that knowledge consists: there must be an observable difference between the behavior or capacities of someone who is said to have that knowledge and someone who is said to lack it. Hence it follows, once more, that a grasp of the meaning of a ... statement must, in general, consist of a capacity to use that statement in a certain way, or to respond in a certain way to its use by others.

In the first part of this passage, Dummett invokes infinite-regress considerations to establish that knowledge of meaning is not "in general" verbalizable, and even that it is "in general" *un*verbalizable. If one tries to restate the argument without the use of the puzzling phrase "in general," then one finds that all the infinite-regress considerations seem to establish is that for *the part of language learned first*, the most elementary part, knowledge of meaning is unverbalizable. It would then seem that any behavioristic conclusions drawn from the argument as a whole ought to be restricted to this part of language: nothing follows about the part of language learned later, the more advanced part.[5]

In the second part of the passage under examination, Dummett invokes a premise about unverbalizable knowledge to reach a conclusion about knowledge of meaning. To avoid equivocation, we must distinguish four claims here, some stronger, some weaker, some more general, some more specific:

[5] Paul Benacerraf suggested to me in general terms that Dummett's arguments might have force for one part of language but not another.

(a) Ascriptions of knowledge of meaning must be supported by appeal to observable evidence.
(b) Knowledge of the meaning of an expression consists in no more than the ability to use it in a certain way.
(c) Ascriptions of implicit knowledge must be supported by appeal to observable evidence.
(d) Implicit knowledge consists in no more than the ability to behave in a certain way.

The anti-behaviorist rejects (b) but accepts (a). (The anti-behaviorist account of the observable evidence supporting an ascription of knowledge of meaning has been reviewed schematically above.) Since the conclusion Dummett desires is (b) and not (a), the premise he requires is (d) and not (c), even if his own formulations are less than unequivocal (1973a, p. 217; 1977, p. 376):

> Implicit knowledge cannot, however, meaningfully be ascribed unless it is possible to say in what the manifestation of that knowledge consists . . .
>
> [A]n ascription of implicit knowledge must always be explainable in terms of what counts as a manifestation of that knowledge, namely the possession of some practical ability.

The anti-behaviorist will argue that if – as the implanted-computer example suggests – (b) is false, then this implies that (d) is false. Dummett, however, invokes the controversial premise (d) without supporting considerations, as if it were self-evident. For this reason anti-behaviorists may with some justice reject the "manifestation" argument as manifestly question-begging.

Circularities in Dummett's arguments for behaviorism do not, however, deprive his case against verism of all its force. Dummett's complaint against verism comes down to this, that verists have returned no answer, formulated in behavioral terms, to the following question: in what can a grasp of the correctness conditions for a sentence consist if the correctness of that sentence need not even in principle be potentially recognizable by human beings? So long as one insists that verists must return an answer, formulated in *psychological* terms, to the foregoing question, one will have to sympathize with Dummett's complaint against verism, even if one does not sympathize with his behaviorism, and adopts the approach of introspective or of physiological rather than behavioral psychology. For the best-known advocates of verism either return no answer at all to Dummett's question; or worse, they answer that to grasp the truth conditions of a sentence is to associate with that sentence the set of *possible worlds* where it is true, but do not explain how a mind or brain confined to the actual world can effect such an association.

One might also sympathize with Dummett's rejection of verism for any of a number of reasons quite unlike Dummett's own, for example, on account of the apparent conflict between truth-conditional theories of meaning and theories of truth in the style of Tarski or Kripke. This route to a rejection of verism is worth mentioning here because Dummett himself sometimes touches on it tangentially in his writings. In one paper (Dummett 1959) he notes that there appears to be a conflict between the view that a biconditional like (2) above constitutes an account of the meaning of "or" and the view that it constitutes part of an account of the meaning of "true," a point that Tarski has also discussed in one of his papers (1944) as an anonymous objection against his theory. Truth-conditional theories of meaning appear to regard truth as a primitive concept, possession of which is a prerequisite for any language-learning, while a theory like Kripke's appears to regard the concept of truth as one acquired fairly late in the process of language-learning, when the learner has acquired a fairly extensive ability to talk of persons, places, and things, and is beginning to learn to talk of talk.[6]

For any of a number of reasons, good or bad, like or unlike Dummett's, many philosophers of language now reject verism. In inveighing against verism, Dummett is to a large extent preaching to the converted. Dummett himself recognizes that Wittgenstein, for one, and Quine, for another, have rejected verism. Yet somewhat surprisingly he can be found writing: "[T]he idea that a grasp of meaning consists in a grasp of truth-conditions was [in 1959] and still is [in 1978], part of the received wisdom among philosophers" (1978, p. xxi). A poll of my own department convinces me that Dummett is wrong here.[7] How could a theory rejected by a constellation of such luminaries as Wittgenstein, Tarski, Quine, Kripke, and Harman, not

[6] A not unrelated reason for rejecting truth-conditional theories of meaning is advanced in Harman (1982):

> Davidson, Lewis, and others have argued that an account of the truth conditions of sentences of a language can serve as an account of the meanings of those sentences. But this seems wrong. Of course, if you know the meaning in your language of the sentence S, and you know what the word "true" means, you will also know something of the form "S is true if and only if . . ."; for example, "'Snow is white' is true if and only if snow is white" or "'I am sick' is true if and only if the speaker is sick at the time of utterance". But this is a trivial point about the meaning of "true", not a deep point about meaning.

For more on the philosophical significance (or lack of it) of the concept of truth, see Soames (1984).

[7] Saul Kripke has suggested that Dummett's statement may be accurate as an account of local conditions at Oxford. But surely it would bespeak a certain parochialism to confuse "fashionable among Oxford philosophers" with "received among philosophers generally."

to mention Dummett himself, be considered "received wisdom among philosophers"?

For many philosophers what is most puzzling about Dummett's case for intuitionism will not be the question arising in his case against Platonism:

(a) Why are we supposed to reject verist semantics?

but rather the question arising in his case against formalism:

(b) How is the rejection of verist semantics supposed to lead to the rejection of classical mathematics?

This latter question will now be taken up.

3 DUMMETT'S CASE AGAINST FORMALISM

Criticism of a theory of language may take any of three forms. The *ordinary descriptive* critic advances evidence that we do not actually speak a language of the sort the theory depicts, whether or not we ought to. The *radical descriptive* critic advances evidence that we could not possible speak a language of that sort, so that the question whether we ought to do does not arise. The *prescriptive* critic (or advocate) advances motives why we ideally ought not (or ought) to speak such a language, whether or not we currently do.

Dummett is a philosopher not primarily renowned for the clarity of this prose, and the interpretation of his works will always be a matter of controversy, not least because he declines to distinguish explicitly factual or descriptive from normative or prescriptive considerations. As I interpreted it in §2, Dummett's criticism of Platonist or verist theories of meaning was descriptive and radical, claiming that we could not possibly possess (because we could not possibly acquire or manifest) a grasp of correctness conditions that are transcendent rather than immanent. As I interpret it, Dummett's advocacy of intuitionist or verificationist theories of meaning is prescriptive. For surely he cannot claim that such theories describe and explain the actual, current patterns of usage of any but a tiny minority of (Dutch) mathematicians. As I interpret it, Dummett's position is that no theory of language of the first type, verist or verificationist, provides a description and explanation of the actual, current patterns of use of the overwhelming majority of mathematicians.

Looking beyond theories of the first type, especially important for Dummett are what I will call neutrally *theories of language of the second type* or *dualist* theories. Such theories depict language as containing two different kinds of sentences: *primary* sentences, possessing *decidable* correctness conditions, and *secondary* sentences, lacking correctness conditions.

The best-known theories of this type are the theory of mathematical language associated with the name of Hilbert (1925), and the theory of scientific language associated with the name of Quine (1951b). On the former theory, primary sentences, called *inhaltlich*, consist of simple arithmetical sentences decidable by computation; secondary sentences, called *ideal*, may contain non-computational mathematical vocabulary (e.g. that of set theory). On the latter theory, primary sentences, said to "lie on the periphery," consist of simple empirical sentences decidable by observation; secondary sentences, said to "lie in the interior," may contain non-observational scientific vocabulary (e.g. that of quantum theory). Hilbert's overall position in philosophy of mathematics is usually called *formalism*. Quine's overall position in philosophy of science is usually called *holism*. Both labels have been used in the literature with so many different connotations that they are perhaps best avoided. Dummett repeatedly stresses the affinities between Hilbert and Quine (1973a, p. 219; 1977, p. 397).[8]

On either theory, primary sentences are distinguished from secondary sentences by their restricted vocabulary. Not only is their non-logical, mathematical or scientific, vocabulary restricted to be computational or observational, but also their logical vocabulary is restricted. According to the precise version of the theory being considered, primary sentences are required to be either *atomic*, containing no logical particles at all, or else to be *quantifier-free*, containing only connectives.

On either theory, secondary sentences serve merely as intra-linguistic instruments for deducing primary sentences, and not as representations of any extra-linguistic reality. Computational and observational facts are represented by primary sentences. It is claimed that the scope, accuracy, and efficiency of the representation of computational and observational facts is enhanced by the presence in the language of sentences that do not themselves represent such facts but that can be used to deduce sentences that do.

On either theory, the question arises how the ability to use secondary sentences can be learned. For example, how is the ability to use disjunctions

[8] It is something of an oversimplification to describe Quine as a dualist, inasmuch as he often indicates that he regards the distinction between the observational periphery and the theoretical interior as a matter of degree rather than kind. But for Dummett the similarities between Quine's position and that of the prototypical dualist Hilbert are more important than such differences.

in deductions acquired? On a verist or verificationist theory the answer is: by grasping the correctness conditions (1) or (2) above. This answer is not available on a dualist theory, and no explicit answer is offered in Hilbert (1925) or Quine (1951b). There is, however, an answer that immediately suggests itself, namely, that the ability is acquired by directly grasping such rules of implication as the following:

(3) A disjunction is implied by each of its disjuncts.

A disjunction implies whatever is implied by each of its disjuncts. In other words, the "meaning of the logical constants" – if what determines their use may be called their "meaning" – consists "directly in the validity or invalidity of possible forms of inference" (Dummett 1977, p. 363). It seems to be this answer that Dummett associates with dualism. It is worth mentioning that quite apart from any general dualist views, the specific view that an account of the "meaning" of the logical particles is best given in terms of such implication conditions as (3) rather than such truth conditions as (2) has had many distinguished advocates, including (according to Prior 1960) several of Dummett's Oxford colleagues.

Dualist theories may be called *semi-verificationist.* They are not verificationist in the strict sense, since some sentences are not assigned correctness conditions. They are verificationist in a loose sense, since all sentences that *are* assigned correctness conditions are assigned decidable, recognizable, verifiable correctness conditions.

May dualist theories of language be called theories of *meaning*? Dummett sometimes takes "(theory of) meaning" in a broad sense, and insists on an affirmative answer (1973c, p. 378):

> A model of language may also be called a model of meaning, and the importance of the conception of language sketched at the end of "Two Dogmas" was that it gave in succinct form the outline of a new model of meaning. It is well known that some disciples of Quine have heralded his work as allowing us to dispense with the notion of meaning. But even the most radical of such disciples can hardly propose that we may dispense with the notion of knowing, or having mastery of, a language; and there is nothing more that we can require of a theory of meaning than that it give an account of what someone knows when he knows a language . . . [W]hatever warrant there may be for asserting that Quine has destroyed the concept of meaning does not appear from the "Two Dogmas" model of language taken by itself. That has merely the shape of one theory or model of meaning among other possible ones.

However, Dummett sometimes takes "(theory of) meaning" in a narrow sense, and insists on a negative answer (1973b, p. 309):

The theory of meaning, which lies at the foundation of the whole of philosophy, attempts to explain the way in which we contrive to represent reality by means of language. It does so by giving a model for the content of a sentence, its representative power. Holism is not, in this sense, a theory of meaning: it is the denial that a theory of meaning is possible.

When in the narrow, negative mood (as throughout 1977) Dummett is prepared to join Brouwer and Heyting in declaring that many of the sentences of classical mathematics are "incoherent" and "unintelligible." This sounds odd. For Dummett can hardly deny that the sentences of classical mathematics possess a definite *usage* within pure mathematics and a definite *utility* through applied mathematics. How can he, as a professed adherent of the slogan that meaning is use, then deny that those sentences have a meaning? Taken literally, the slogan implies that a sentence having a use *thereby* has a meaning. The answer, the explanation of the oddity, is, of course, that Dummett, as we have already seen, adheres to the slogan that meaning is use only in a non-literal, almost idiosyncratic, sense.

The narrow, negative terminology need not be misleading *provided the following point is never forgotten*: When Dummett says that many sentences of classical mathematics lack meaning-in-the-narrow-sense, he is *only* saying (in highly emotive terms) that the theory of meaning-as-conditions-for-correctness-as-a-representation-of-reality is inapplicable to those sentences. This *factual* claim about how language *is* cannot by itself imply any *normative* claim about how mathematics *ought to be*. Some extra, tacit premise of a normative or prescriptive character is needed. As I interpret it, Dummett's criticism of dualism is prescriptive, and rests on an extra, tacit premise of *anti-instrumentalism* or *representationalism*, according to which every sentence of a language ideally ought to play a representational rather than a merely instrumental role. Thus when Dummett writes: "A sentence is a representation of some facet of reality" (1973b, p. 309), according to my interpretation he has not quite accurately reflected his own view: "ought to be" ought to be where "is" is in the quoted formulation.

The requirement of representationality is, of course, accepted by Platonists, who hold, in opposition to intuitionists and formalists, that this requirement is *already* met by our current language. Representationalism unites Platonists and intuitionists in opposition to formalists, much as behaviorism unites Quineans and Dummettians in opposition to Davidsonians and Chomskians. (There are, however, important differences between the Harvard behaviorism of Skinner or Quine and the Oxford behaviorism of Ryle or Dummett. Moreover, it is not obvious that a verificationist or dualist *must* be a behaviorist.)

Consider the situation of a philosopher initially sympathetic, for behavioristic or other reasons, to a naive descriptive verificationism like that of the early positivists, who comes to appreciate that such a theory is inadequate as an account of the actual, current patterns of use in our language. One response would be to revise the theory to fit the facts of language, perhaps falling back to a semi-verificationist, dualist position. Another response would be to require a revision of language, to fit the norms of the theory. Quine and Dummett exemplify these two responses. Dummett's against Quine stands or falls with the success or failure of his attempts to motivate the requirement of representationality.

In both the texts under examination Dummett discusses, by way of offering such motivation, the following worry about languages of the sort depicted by dualist theories: in such a language, there is a threat of deducing *incorrect* primary sentences by means of secondary sentences. In one text, the worry seems to be that incorrect primary sentences might be deduced *from* (theories composed of) secondary sentences (1973a, p. 220):

> With what right do we feel assurance that the observational statements deduced with the help of complex theories, mathematical, scientific and otherwise, embedded in the interior of the total linguistic structure, are true, when these observation statements are interpreted in terms of their stimulus meanings? To this the holist attempts no answer, save a generalised appeal to induction: these theories have "worked" in the past, in the sense of having for the most part yielded true observation statements, and so we have confidence that they will continue to work in the future.

This worry, or rather, the demand for a guarantee against it, is easily dismissed. Of course there is a threat that a scientific theory about, say, black holes or quarks may have incorrect observational consequences. We have seen such threats realized many times in the history of science, and we have known since the time of Hume that there can be no guarantee against them. And of course there is a threat that a mathematical theory about, say, *l*-adic cohomology or ω-complete ultrafilters may have incorrect and even inconsistent computational consequences. We have seen such threats realized a few times in the history of mathematics (in connection with infinitesimal calculus and naive set theory), and we have known since the work of Gödel that there can be no guarantee against them. If and when such threats are again realized, we will, as we always have in the past on such occasions, revise our *theories*. But why even then, let alone now, revise our *logic*? It is a delusion to imagine that a preemptive change of logic could provide a guarantee against such threats, unless, indeed, the new logic were so restrictive as to make the formulation of any non-trivial theories impossible.

In the other text, the worry seems to be that incorrect primary sentences might be deduced from *correct* primary sentences *by way of secondary sentences*. On a *sequential* formulation of logic, this is the worry that a sequent

(*) $A_1, \ldots, A_n \Rightarrow B$

with the A_i primary and correct and B primary and incorrect, might be deducible by pure classical logic, if secondary sentences are allowed to appear in the deduction. *If only primary sentences are allowed to appear in the deduction, there is nothing to worry about,* since primary sentences are decidable, and not even intuitionists doubt the trustworthiness of classical logic as applied to decidable sentences. As Dummett says, it would be a "severe defect" in the classical rules of implication if by means of them "we can construct a deductive chain leading from correct premises to an incorrect conclusion" (1977, p. 364). He reminds us that even on a dualist theory there is an external standard against which the acceptability of rules of implication is to be judge. A rule is acceptable only if it is *sound*, only if it preserves correctness *in all instances where the notion of correctness is applicable*, that is, in all instances where the premises and conclusion are all primary sentences. (Thus even if the "meaning" of the logical particles is given by implication conditions, not just any old particles and conditions will do. This point has been illustrated by Prior (1960). Dummett does not claim that classical logic is, in this sense, *demonstrably unsound* (as is Prior's "tonk" logic). What worries Dummett is that classical logic seems to be *not demonstrably sound.* He desires a guarantee of soundness, or what would, as we have seen above, be sufficient for this, a guarantee of *conservativeness*, a guarantee that the addition of the secondary sentences to the language does not permit the deduction of any sequences (*) involving only primary sentences that were not deducible already (1977, pp. 363–4).

Dummett seems to hold that such a guarantee could only be provided by a *semantic* soundness or conservativeness proof, and that such a proof or "justification" will be available only if we revise the language and extend the assignment-of-correctness-conditions or "interpretation" to "all statements or formulas with which we are concerned" (1977, p. 220). Dummett seems to overlook the possibility of a purely *syntactic* proof of soundness or conservativeness. As Richard Grandy has pointed out in a perceptive review (Grandy 1982), just such a guarantee as Dummett seems to desire is provided by the famous Cut-elimination Theorem of Gentzen, according to which any sequent (*) that has a deduction at all has a deduction in which no symbols occur that do not occur in (*) already. Moreover, though Gentzen's theorem is about *classical* logic, Gentzen's proof is given in *intuitionistic* metamathematics. Thus the threat that worries Dummett seems elusive, to say the least.

In any case, the guarantee he desires would be intangible, on his own admission. For it is precisely the theme of Dummett (1973b) that no "justification of deduction" or soundness of proof can be "suasive," that is, can persuade anyone sincerely in doubt as to the soundness of the logic. For any such proof, being a proof, would itself use logic.

In opposition to Kreisel, Dummett is concerned to argue for intuitionism not as one legitimate form of mathematics among others, but as the *sole* legitimate form (1977, p. 360). Dummett is concerned to argue for a revision amounting not to a reform, but to a revolution, in mathematics. Any revolution involves costs that the benefit of an intangible guarantee against an elusive threat of unsoundness seems insufficient to outweigh. It seems that its desirability as a means toward the end of guaranteeing soundness is not a consideration sufficient to motivate the requirement of representationality.

Dummett might, of course, rest his case against formalism on the desirability of representationality as an end in itself. That he indeed values representationality highly for its own sake is suggested by his applying the uncalled-for emotive term "unintelligible" to sentences that he knows perfectly well how to use, but that happen to lack meaning-in-the-narrow-sense-of-conditions-for-correctness-as-a-representation-of-reality. Dummett's value judgment might, however, be questioned by many mathematicians.

It hardly needs saying that the requirement of representationality will be rejected by the many pure mathematicians who value mathematics as an art. From their point of view, there is no sole legitimate form of mathematics. A mathematician may work now in intuitionistic, now in classical mathematics, just as a painter may work now in a representational, now in an abstract style. Personal taste will dictate how much time is devoted to each, though it may be said that the overwhelming majority of mathematicians find more beauty in the classical than in the intuitionistic style.

What does perhaps need saying is that the requirement of representationality may also be questioned by the many applied mathematicians who value mathematics for its contribution, through science, to the theoretical prediction and practical control of experience. From their point of view, it is essential that language should contain *some* sentences to serve as records or predictions of experience, as representations of empirical reality. But beyond this it seems wisest to accept the advice of Carnap (1950) and be "tolerant in permitting linguistic forms." It is questionable whether the scope, accuracy, and efficiency of applications to the empirical world would be enhanced by imposing the restriction that *all* sentences must play a representational rather than a merely instrumental role in language. Many physicists, mathematicians, logicians, and philosophers have suggested precisely the contrary: that intuitionistic restrictions on mathematics would be detrimental to applications. Such views as the following are often voiced (Manin 1977, pp. 172–3):

> [C]onstructivism is in no sense "another mathematics". It is, rather, a sophisticated subsystem of classical mathematics, which rejects the extremes in classical mathematics, and carefully nourishes its effective computational apparatus. Unfortunately, it seems that it is these "extremes" – bold extrapolations, abstractions which are infinite and do not lend themselves to a constructive interpretation – which make classical mathematics effective. One should try to imagine how much help mathematics could have provided twentieth century quantum physics if for the past hundred years it had developed using only abstractions from "constructive objects."

I do not pretend to be an expert in such matters, but there are several studies in the literature that seem to me to indicate that such complaints are not entirely without foundation. As one example, there is an important series of papers by Pour-El and Richards (1979–87) establishing that much of the machinery of functional analysis deployed in quantum physics cannot be developed in its usual form *with recursive analysis*. And experience shows that what can or cannot be done recursively is a usually reliable (though by no means infallible) guide to what can or cannot be done intuitionistically.

As another example, there is a paper of Douglas Bridges (1981), examining quantum physics from an intuitionistic viewpoint. Bridges is, to be sure, a follower of Bishop's *intuitionism-without-choice-sequences* rather than of Brouwer's intuitionism-with-choice-sequences, but Bishop's school has thus far been able to go further than Brouwer's school in reconstructing applicable portions of functional analysis. Bridges is obliged to concede that "a constructive examination of the mathematical foundations of quantum physics does reveal substantial problems." It is also worth mentioning that even if the indications cited are misleading and it turns out to be *possible in principle* to get by with intuitionistic functional analysis in applications to quantum physics, getting by in this way is very likely to be *infeasible in practice*.

Table 14.1

Philosophies of mathematics	
PLATONISM	FORMALISM (Hilbert)
Associated theories of meaning	
VERISM	HOLISM (Quine)
Character of objection to theory of meaning	
RADICAL DESCRIPTIVE	PRESCRIPTIVE
We do not and could not speak a language of the sort described by the theory	We ought not to speak a language of the sort described by the theory
Principle on which the objection is based	
BEHAVIORISM	REPRESENTATIONALISM
There can be nothing more to knowing the meaning of a sentence than being able to use it	Every sentence ought to serve as a representation of extra-linguistic reality, not a mere intra-linguistic instrument for deducing other sentences
Argument for principle	
If there were anything more to such knowledge, the extra ingredient could be neither acquired nor manifested.	Representationality is desirable (a) as a means towards guaranteeing soundness (b) as an end in itself
Comment	
The acquisition and manifestation arguments involve fallacies and circularities	(a) Soundness can be guaranteed without representationality (b) Representationality may conflict with the desirable end of applicability

Whether applications to the empirical world are of value is a question on which philosophers' judgments vary over a wide spectrum. On the far right stands Plato, who regarded such applications as evil. On the far left stand the Gang of Four, who regarded the development of mathematics for any purpose but such applications as evil. Among intuitionists, Brouwer was, in this respect, a thoroughgoing Platonist, with an attitude not of passive indifference, but of active hostility towards applications (see van Stigt 1979). Weyl, however, seems to have been uneasy over his inability to reconcile his philosophical attraction towards intuitionism with his scientific interest in applications.

One need not be a Maoist to sympathize with this unease, and to be disturbed by an argument for the claim that intuitionism is the sole legitimate form of mathematics in which any consideration of widely held doubts as to the adequacy of intuitionism for applications is omitted. (The omission is the more surprising in an argument directed against *Quine*, since doubts as to adequacy for applications have been central among Quine's objections against other revisionist proposals such as those of the nominalists.) The omission suggests a tacit system of values so unworldly as to be *irresponsible*.

Dummett may perhaps be absolved personally from charges of irresponsibility and inquisitorial interference with science. For though he advances an argument for intuitionistic revisionism, he is cautious enough to distance himself personally somewhat from that argument. He is not so bold as to claim that his conclusion ought to be accepted and put into practice. What he claims is that it is an argument "of considerable power" (1973a, p. 226). In view of the gaps and weaknesses in the argument that I have tried to point out, even this more cautious claim might well be challenged.

4 SUMMARY

Table 14.1 summarizes my interpretation of and commentary on Dummett's case for intuitionism.

Annotated bibliography

TECHNICAL PAPERS ON SET THEORY

"Forcing" (Burgess 1977a)

The continuum hypothesis (CH) states that the continuum, the cardinal of the set of real numbers, is equal to aleph-one, the least cardinal greater than that of the set of natural numbers. The two great methods for proving the consistency and independence relative to the usual Zermelo–Frankel axioms of set theory are the method of inner models, exemplified by Gödel's constructible sets, through which he proved the consistency of CH, and the method of forcing, through which Paul Cohen proved the independence of CH. Burgess (1977a) is an exposition for forcing intended to make the method available for use by non-specialists, by following Robert Solovay and reducing what needs to be understood about forcing in order to apply the method to three "axioms of forcing" whose proof can be left to specialists.

"Consistency proofs in model theory" (Burgess 1978a)

Just as ordinary arithmetic, the theory of addition and multiplication, has an analogue in transfinite cardinal and ordinal arithmetics, so ordinary combinatorics, the theory of permutations and combinations, has a transfinite analogue in combinatorial set theory. Generalized-quantifier logic is a family of extensions of first-order logic that retains the same notion of model, but adds additional clauses to the definition of truth-in-a-model to cover such operators as "there exist infinitely many" or "there exist uncountably many." Hypotheses in combinatorial set theory turn out to have implications for generalized-quantifier logic. Ronald Jensen, through a deep analysis of the fine structure of Gödel's constructible sets, proved the consistency of many combinatorial principles, and therewith of certain principles about generalized-quantifier logic implied by them. This paper shows how one of Jensen's deepest consistency results about generalized-quantifier logic can be obtained more easily using the method of forcing.

"Descriptive set theory and infinitary languages" (Burgess 1977b)

Descriptive set theory is the branch concerned with definable sets of real numbers or linear points. Infinitary logic is a family of extensions of first-order logic that retains the same notion of model, but adds additional clauses to the definition of truth-in-a-model to cover such operations as the conjunction and disjunction of infinite sets of formulas. Several workers, most notably Robert Vaught, showed the applicability of some results in descriptive set theory to infinitary logic. This paper presents my contributions to Vaught's project, the main one being as follows. Jon Barwise developed a notion of "absoluteness" for logics, and characterized the logics absolute relative to the weak Kripke–Platek set theory as the sublogics of a certain well-known infinitary logic. This paper characterizes the logics absolute relative to standard Zermelo–Frankel set theory, essentially as those whose sentences admit suitable "approximations" by sentences of that same infinitary logic.

"Equivalence relations generated by families of Borel sets" (Burgess 1978b)
"A reflection phenomenon in descriptive set theory" (Burgess 1979a)
"Effective enumeration of classes in a Σ^1_1 equivalence relation" (Burgess 1979b)

The special sets of real numbers (or pairs of real numbers) that are studied in descriptive set theory are called *projective*, and they are divided into several classes of increasing complexity and diminishing tractability: Borel or Δ^1_1, analytic or Σ^1_1, coanalytic or Π^1_1, Δ^1_2, Σ^1_2, Π^1_2, and so on. A basic question of set theory is how many elements a set of real numbers can have. For projective sets up to analytic it can be proved that they contain either countably many or perfectly many (a stronger condition implying but not implied by continuum many); for higher projective sets the same result requires large cardinal axioms going beyond Zermelo–Frankel set theory. A related question is how many equivalence classes an equivalence relation that is projective (considered as a set of pairs of real numbers) can have. Jack Silver proved that the answer is countably or perfectly many if the equivalence relation is coanalytic. It had long been known that if the equivalence relation is analytic there is a third possibility: aleph-one many but not perfectly many. Harvey Friedman asked whether these are the only possibilities. The first two papers listed above together supply the affirmative answer that was the main result of my doctoral dissertation, written under Silver's direction, while the third offers a refinement. The first shows that any analytic equivalence relation is the intersection of aleph-one Borel equivalence relations, and the second uses Silver's theorem to show that the intersection of aleph-one Borel equivalence relations has countably many, aleph-one many, or perfectly many classes, a result later extended by Saharon Shelah to further projective equivalence relations using large-cardinal assumptions.

"A selection theorem for group actions" (Burgess 1979c) "A measurable selection theorem" (Burgess 1980a) "Sélections mesurables pour relations d'équivalence à classes G_δ" (Burgess 1980b) "Careful choices: a last word on Borel selectors" (Burgess 1981d) "From preference to utility: a problem of descriptive set theory" (Burgess 1985)

Given a function assigning to each real number *x* a set of real numbers *F*(*x*), the axiom of choice (AC) guarantees that there exists a function defined for those *x* for which *F*(*x*) is non-empty, and assigning to each such *x* an element *f*(*x*) of *F*(*x*). But AC does not guarantee that this *f* has any nice properties of the kind one studies in calculus, such as differentiability or integrability. Measurable selection theorems are results to the effect that, assuming certain niceness conditions on *F*, there follows the existence of an *f* with certain corresponding niceness conditions. The first four papers in this group provide three examples, differing in their niceness assumptions about *F* and niceness conclusions about *f*. Measurable selection theorems are used in what I have elsewhere called "the (hyper)theoretical fringes of subjects whose core is applied." The last paper in the group uses one to solve a problem raised by R. D. Mauldin in mathematical economics.

"What are R-sets?" (Burgess 1982a) "Classical hierarchies from a modern standpoint, parts I & II" (Burgess 1983a and b)

One of the main tools in modern descriptive set theory is the "game quantifier" of Yannis Moschovakis (which is also, to allude back to the discussion of infinitary logic above, the basis for the most important logic that is Zermelo–Frankel but not Kripke–Platek absolute). It is known that, in a sense that can be made precise, application of this quantifier to Borel sets yields Δ^1_2 sets. Borel sets themselves are divided into classes of increasing complexity: closed or Π^0_1 sets, open or Σ^0_1 sets, Δ^0_2 sets, Π^0_2 sets, Σ^0_2 sets, Δ^0_3 sets, and so on. Moschovakis had already established that applying the game quantifier to closed and open sets yields analytic and coanalytic sets, respectively. The first paper listed above is a semi-popular introduction to the pair of papers listed below it, which show that from Δ^0_2 and Δ^0_3 sets one obtains, respectively, families of sets studied during the 1920s and 1930s under the names of C-sets and R-sets.

"The truth is never simple" (Burgess 1986)
"Addendum to 'The truth is never simple'" (Burgess 1988)

The first paper is concerned with determining the place in the classifications of descriptive set theorists of the set of all (Gödel numbers of) truths of arithmetic according to various recent theories of truth: four versions of Kripke's view (with the Kleene three-valued or with van Fraassen supervaluation schemes for handling truth-value gaps, and with the minimal fixed point or with the maximal intrinsic fixed point) and three versions of the revision view (Gupta's, Hertzberger's, and Belnap's). A complete set of answers is obtained except for one case that had to wait for the second paper, and one other case that my work left open. In the course of working out the answer, examples are provided of sentences that are true on some of the views but not on the others; indeed, examples of all possible combinations are given.

"Sets and point-sets" (Burgess 1990)

Perhaps the most surprising discoveries to emerge from foundational work in the last century are two. First, all the natural choices turn out to be, for reasons for which we at present lack any clear explanation, linearly ordered in strength (for the cognoscenti, I mean in consistency strength), so that given any two choices, it always turns out that one is stronger than the other, unless they turn out to be, despite superficial differences, of equal strength. Second, stronger and stronger axioms, though about objects of higher and higher level or type, continue to have more and more implications even about objects of the very lowest level and type. What this latter cryptic assertion means is spelled out in the case of *geometrical* theories in this semi-technical, largely expository paper. The aim is to explode the myths, which have gained some currency among philosophers not well-trained in logic, that "mathematics is conservative over physics" and that "higher mathematics is conservative over lower mathematics."

"A remark on Henkin sentences and their contraries" (Burgess 2003b)

Jaakko Hintikka has for some time now been advocating what he has called IF (for "independence-friendly" or "information-friendly") logic. This logic is "non-classical," not in the way that modal and tense and probability and conditional and intuitionistic logic are, requiring a different notion of model from that used in first-order logic, but rather in the way that generalized-quantifier and infinitary logic are. The logic has been found interesting from a technical point of view even by many who have not found Hintikka's large philosophical claims for it convincing. The paper establishes a technical result to the effect that the logic admits no semantic operation of negation, a result that it is hard to interpret as having any but negative implications about the philosophical claims for the logic, though this point is not strenuously argued in the paper.

TECHNICAL PAPERS ON TENSE AND OTHER NON-CLASSICAL LOGICS

"Basic tense logic" (Burgess 1984a)

Basic tense logic adds to classical sentential logic new one-place connectives P for "it was the case that" and F for "it will be the case that." Soundness and completeness theorems establish that different sets of axioms for P and F exactly correspond to different assumptions about the structure of time. (Is it linearly ordered? Does it have a first or last moment? Are the moments of time densely ordered or discrete?) When I first read the statements of a group of such theorems, I worked out for myself proofs based on what has come to be called the "step-by-step" method, and erroneously concluded that this must be the method everyone was using. In fact, only a few were, while most were using Segerberg's "bull-dozing" and "unraveling" methods instead; but my adoption of the step-by-step method in this expository chapter helped popularize it, first among philosophical logicians, then among theoretical computer scientists.

"The unreal future" (Burgess 1979d)
"Decidability and branching time" (Burgess 1980c)

These two were originally a single item. Division was suggested by Krister Segerberg, regular editor of *Theoria* and guest editor of *Studia Logica*. One of the first applications of modern tense logic envisioned by its founder, Arthur Prior, was to the analysis of traditional debates over future contingents, and of the underlying picture of a time in which the one past behind us branches into many possible futures before us. As part of this analysis Prior distinguished two positions called, after two historical figures, "Peircean" and "Ockhamist." The former does not, while the latter does, consider it meaningful to speak, even now, of some unknown one from among the many possible futures as being the future that will become actual. The first section of "The unreal future" offers a more formal account of Prior's conception of the interaction between temporal and modal operators than he himself gave, then gives an informal summary of the formal results in "Decidability and branching time," which shows that the set of sentences valid in all Peircean models is decidable, while for Ockhamism one must distinguish standard from non-standard models. The second section of the same paper was ostensibly a survey of conceptual issues that needed to be addressed before moving beyond the sentential to a predicate logic of branching time, but was really my first attempt to come to grips with "Naming and necessity." Underlying it was, in embryonic and inchoate form, the thought that "metaphysical" modality might be demystified by tracing its source to *our* sortal classifications of the objects of our thought, and *our* conventions as to what does and what does not count as another appearance of the same thing of a given sort. The idea was not successfully worked out in the paper, and I have not as yet even today achieved a development of it that is wholly satisfactory even to myself, though the idea still seems to me to be promising.

"Axioms for tense logic" (Burgess 1982b)

The more important first part of this two-part paper concerns the two-place "since" and "until" operators by which Hans Kamp enriched Prior's tense logic. The paper provides an axiomatization proved sound and complete by an extension of the step-by-step method.

"The decision problem for linear temporal logic" (Burgess and Gurevich 1985)

Segerberg's "filtration" method can be used to prove the decidability of the tense logics appropriate to many different models of time, but not the model on which the moments of time are ordered like the real numbers. This paper presents two different proofs of the decidability of that logic, along quite different lines. The first is mine. Gurevich, the referee for the paper in which I wrote up this method, suggested the second, using more advanced techniques and thereby providing some additional information. We agreed it was best for him to become a co-author, so that both methods could be made available in a single paper.

"Probability logic" (Burgess 1969)

This paper provides a complete axiomatization and a decision theory for a sentential modal logic enriched with an operator "it is probable that" in addition to the operator "it is necessary that." The axioms are those appropriate for a qualitative notion of probability, though results are also presented on quantitative notions.

"Quick completeness proofs for some logics of conditionals" (Burgess 1981b)

Robert Stalnaker and David Lewis independently developed similar but not identical views on counterfactual or subjunctive conditionals, and there is a third variant as well. This paper provides complete axiomatizations and therewith proofs of decidability for a range of variants. The systems have since been found to admit another interpretation, in terms of nonmonotonic logics of the kind pioneered by Kraus, Lehmann, and Magidor.

"The completeness of intuitionistic propositional calculus for its intended interpretation" (Burgess 1981a)

The question addressed here is whether Heyting's axiomatization of intuitionistic logic is complete in the sense that every sentential formula all of whose instances (obtained by substituting formulas of intuitionistic arithmetic or analysis for its sentential variables) are intuitionistically correct is a thesis of the system. Such a question is *not* answered merely by a proof of completeness for some formal "semantics" on the order of topological models or Kripke models. Georg Kreisel was able, using topological models as a *starting point*, to obtain an affirmative answer, making certain assumptions about intuitionistic analysis (the theory of "lawless" sequences). The paper shows how, on the same assumptions, to obtain an affirmative answer taking Kripke models as a starting point.

References

Ackermann, Diana [Ackermann, Felicia Nimue] (1978) "*De re* propositional attitudes toward integers," *Southwestern Journal of Philosophy* vol. 9, pp. 145–53.

Anderson, Alan Ross and Belnap, Nuel D., Jr. (1975) *Entailment: The Logic of Relevance and Necessity* vol. I (Princeton, NJ: Princeton University Press).

Anderson, C. A. and Zelëny, M. (2002) (eds.) *Logic, Meaning, and Computation: Essays in Memory of Alonzo Church* (Dordrecht: Kluwer).

Azzouni, Jodi (2004) *Deflating Existential Consequence: A Case for Nominalism* (Oxford: Oxford University Press).

Baire, René, Borel, Émil, Hadamard, Jacques, and Lebesgue, Henri (1905) "Cinque lettres sur la théorie des ensembles," *Bulletin de la Société Mathématique de France* vol. 33, pp. 261–73.

Balaguer, Mark (1998) *Platonism and Anti-Platonism in Mathematics* (Oxford: Oxford University Press).

Barcan, Ruth C. [Marcus, Ruth Barcan] (1946) "A functional calculus of first order based on strict implication," *Journal of Symbolic Logic* vol. 11, pp. 1–16.

(1947) "Identity of individuals in a strict functional calculus of second order," *Journal of Symbolic Logic* vol. 12, pp. 12–15.

Bar-Hillel, Y., Poznanski, E. I. J., Rabin, M. O., and Robinson, A. (1961) (eds.) *Essays on the Foundations of Mathematics* (Jerusalem: Magnes Press).

Barwise, K. Jon (1977) *Handbook of Mathematical Logic* (Amsterdam: North Holland).

Belnap, Nuel D., Jr. and Green, Mitchell (1994) "The thin red line," in Tomberlin (1994), pp. 365–88.

Benacerraf, Paul and Putnam, Hilary (1964) *Philosophy of Mathematics: Selected Readings* (Englewood Cliffs, NJ: Prentice-Hall).

(1983) *Philosophy of Mathematics: Selected Readings*, 2nd edn (Cambridge: Cambridge University Press).

Bernays, Paul (1961) "Zur Frage der Unendlichkeitsschemata in der axiomatischen Mengenlehre," in Bar-Hillel *et al.* (1961), pp. 3–49.

(1976) "On the problem of schemata of infinity in axiomatic set theory," English translation of Bernays (1961) by J. Bell and M. Plänitz, in Müller (1976), pp. 121–72.

Birkhoff, Garrett (1937) "Rings of sets," *Duke Mathematical Journal* vol. 3, pp. 443–54.
Bochenski, I., Church, A., and Goodman, N. (1956) *The Problem of Universals: A Symposium* (Notre Dame, IN: Notre Dame University Press).
Boolos, George (1984) "To be is to be the value of a variable (or to be some values of some variables)," *Journal of Philosophy* vol. 81, pp. 430–39, reprinted in (Boolos 1997), pp. 54–72.
(1985) "Nominalist Platonism," *Philosophical Review* vol. 94, pp. 327–44, reprinted in Boolos (1997a), pp. 73–87.
(1987) "The consistency of Frege's *Foundations of Arithmetic*," in Thomson (1987), pp. 3–20; reprinted in Demopoulous (1995), pp. 211–33, and in Boolos (1997a), pp. 182–201.
(1993) *The Logic of Provability* (Cambridge: Cambridge University Press).
(1997a) *Logic, Logic, and Logic* (Cambridge, MA: Harvard University Press).
(1997b) "Must we believe in set theory?" in Boolos (1997a), pp. 120–32.
Borges, Jorge Luis (1962) "Tlön, Uqbar, Orbis Tertius," translated from the Spanish by Alastair Reid, in A. Kerrison (ed.) *Ficciones* (New York: Grove Press).
Bridges, Douglas (1981) "Towards a constructive foundation for quantum mechanics," in Richman (1981), pp. 260–73.
Brouwer, L. E. J. (1975) *Collected Works*, vol. I: *Philosophy and Foundations of Mathematics* (Amsterdam: North Holland).
(1976) *Collected Works*, vol. II: *Geometry, Analysis, Topology and Mechanics* (Amsterdam: North Holland).
Bull, R. A. and Segerberg, Krister (1984) "Basic modal logic," in Gabbay and Guenthner (1984), pp. 1–88.
Burgess, John P. (1969) "Probability logic," *Journal of Symbolic Logic* vol. 34, pp. 264–74.
(1977a) "Forcing" in Barwise (1977), pp. 403–52.
(1977b) "Descriptive set theory and infinitary languages," in *Proceedings of the 1977 Belgrade Symposium on Set Theory and Foundations of Mathematics*, Mathematical Institute, Belgrade, pp. 9–30.
(1978a) "Consistency proofs in model theory: a contribution to *Jensenlehre*," *Annals of Mathematical Logic* vol. 14, pp. 1–12.
(1978b) "Equivalence relations generated by families of Borel sets," *American Mathematical Society Proceedings* vol. 69, pp. 323–6.
(1979a) "A reflection phenomenon in descriptive set theory," *Fundamenta Mathematicae* vol. 104, pp. 127–39.
(1979b) "Effective enumeration of classes in a $\sum^1_1$ equivalence relation," *Indiana University Mathematical Journal* vol. 28, pp. 353–64.
(1979c) "A selection theorem for group actions," *Pacific Journal of Mathematics* vol. 80, pp. 333–6.
(1979d) "The unreal future," *Theoria* vol. 44, pp. 157–79.
(1980a) "A measurable selection theorem," *Fundamenta Mathematicae* vol. 100, pp. 91–100.

(1980b) "Sélections mesurables pour relations d'équivalence à classes G_δ," *Bulletin des Sciences Mathématiques* vol. 104, pp. 435–40.

(1980c) "Decidability and branching time," in K. Segerberg (ed.) *Trends in Modal Logic, Studia Logica* vol. 39, pp. 203–18.

(1981a) "The completeness of intuitionistic propositional calculus for its intended interpretation," *Notre Dame Journal of Formal Logic* vol. 22, pp. 17–28.

(1981b) "Quick completeness proofs for some logics of conditionals," *Notre Dame Journal of Formal Logic* vol. 22, pp. 76–84.

(1981c) "Relevance: a fallacy?," *Notre Dame Journal of Formal Logic* vol. 22, pp. 97–104.

(1981d) "Careful choices: a last word on Borel selectors," *Notre Dame Journal of Formal Logic* vol. 22, pp. 219–26.

(1982a) "What are R-sets?" in Metakides (1982), pp. 307–24.

(1982b) "Axioms for tense logic, I. Since and until," *Notre Dame Journal of Formal Logic* vol. 23, pp. 367–74.

(1983a) "Classical hierarchies from a modern standpoint, I. C-sets," *Fundamenta Mathematicae* vol. 115, pp. 81–96.

(1983b) "Classical hierarchies from a modern standpoint, II. R-sets," *Fundamenta Mathematicae* vol. 115, pp. 97–105.

(1983c) "Common sense and 'relevance,'" *Notre Dame Journal of Formal Logic* vol. 24, pp. 41–53.

(1983d) "Why I am not a nominalist," *Notre Dame Journal of Formal Logic* vol. 24, pp. 93–105.

(1984a) "Basic tense logic," in Gabbay and Guenthner (1984), pp. 89–134.

(1984b) "Read on relevance: a rejoinder," *Notre Dame Journal of Formal Logic* vol. 25, pp. 217–23.

(1984c) "Dummett's case for intuitionism," *History and Philosophy of Logic* vol. 5, pp. 177–94.

(1985) "From preference to utility: a problem of descriptive set theory," *Notre Dame Journal of Formal Logic* vol. 26, pp. 106–14.

(1986) "The truth is never simple," *Journal of Symbolic Logic* vol. 51, pp. 663–81.

(1988) "Addendum to 'The truth is never simple,'" *Journal of Symbolic Logic* vol. 53, pp. 390–2.

(1989) "Epistemology and nominalism," in Irvine (1989), pp. 1–15.

(1990) "Sets and point-sets," in Fine and Lepin (1990), pp. 456–63.

(1992) "Proofs about proofs: a defense of classical logic, I," in Detlefsen (1992), pp. 79–82.

(1993) "How foundational work in mathematics can be relevant to philosophy of science," in Hull *et al.* (1993), pp. 433–41.

(1995) "Frege and arbitrary functions." in Demopoulos (1995), pp. 89–107.

(1996) "Marcus, Kripke, and names," *Philosophical Studies* vol. 84, pp. 1–47, reprinted in Humphreys and Fetzer (1998), pp. 89–124.

(1998a) "How not to write history of philosophy," in Humphreys and Fetzer (1998), pp. 125–36.

(1998b) "Occam's razor and scientific method," in Schirn (1998), pp. 195–214.

(1998c) "*Quinus ab omni naevo vindicatus*," in A. A. Kazmi (ed.) *Meaning and Reference: Canadian Journal of Philosophy Supplement* vol. 23, pp. 25–65.

(1999) "Which modal logic is the right one?" *Notre Dame Journal of Formal Logic* vol. 40, pp. 81–93.

(2001) Review of Balaguer (1998), *Philosophical Review*, vol. 101, pp. 79–82.

(2002a) "Nominalist paraphrase and ontological commitment," in Anderson and Zelëny (2002), pp. 429–44.

(2002b) "Is there a problem about deflationary theories of truth?" in Horsten and Halbach (2002), pp. 37–56.

(2003a) "Numbers and ideas," *Richmond Journal of Philosophy* vol. 1, pp. 12–17.

(2003b) "A remark on Henkin sentences and their contraries," *Notre Dame Journal of Formal Logic* vol. 44, pp. 185–8.

(2004a) "Quine, analyticity, and philosophy of mathematics," *Philosophical Quarterly* vol. 54, pp. 38–55.

(2004b) "Mathematics and *Bleak House*," *Philosophia Mathematica* vol. 12, pp. 18–36.

(2004c) "*E pluribus unum*: plural logic and set theory," *Philosophia Mathematica* vol. 12, pp. 193–221.

(2004d) review of Azzouni (2004) *Bulletin of Symbolic Logic* vol. 10, pp. 573–7.

(2005a) "No requirement of relevance," in Shapiro (2005), pp. 727–50.

(2005b) *Fixing Frege* (Princeton, NJ: Princeton University Press).

(2005c) "Translating names," *Analysis* vol. 65, pp. 196–204.

(2005d) "Being explained away," *Harvard Review of Philosophy* vol. 13, pp. 41–56.

(2005e) "On anti-anti-realism," *Facta Philosophica* vol. 7, pp. 121–44.

(forthcoming) "Protocol sentences for lite logicism," in Lindström (forthcoming).

Burgess, John P. and Gurevich, Yuri (1985) "The decision problem for linear temporal logic," *Notre Dame Journal of Formal Logic* vol. 26, pp. 115–28.

Burgess, John P. and Hazen, A. P. (1998) "Arithmetic and predicative logic," *Notre Dame Journal of Formal Logic* vol. 39, pp. 1–17.

Burgess, John P. and Rosen, Gideon (1997) *A Subject With No Object: Strategies for Nominalistic Interpretation of Mathematics* (Oxford: Oxford University Press).

Cantor, Georg (1885) review of Frege (1884), *Deutsche Literaturzeitung*, vol. 6, pp. 728–9.

Carnap, Rudolf (1946) "Modalities and quantification," *Journal of Symbolic Logic* vol. 11, pp. 33–64.

(1947) *Meaning and Necessity: A Study in Semantics and Modal Logic* (Chicago: University of Chicago Press).

(1950) "Empiricism, semantics, and ontology," *Revue Internationale de Philosophie* vol. 4, pp. 20–40.

Chihara, Charles (1973) *Ontology and the Vicious Circle Principle* (Ithaca, NY: Cornell University Press).

(1989) "Tharp's 'Myth and Mathematics,'" *Synthese* vol. 81, pp. 153–65.
(1990) *Constructibility and Mathematical Existence* (Oxford: Oxford University Press).
Chomsky, Noam (1959) review of Skinner (1957) *Language* vol. 35, pp. 26–58.
Church, Alonzo (1950) review of Fitch (1949) *Journal of Symbolic Logic* vol. 15, p. 63.
Cocchiarella, Nino (1984) "Philosophical perspectives on quantification in tense and modal logic," in Gabbay and Guenthner (1984), pp. 309–53.
Cohen, R. S. and Wartofsky, M. W. (1965) (eds.) *Boston Studies in the Philosophy of Science*, vol. II (New York: Humanities Press).
Copeland, B. J. (1979) "When is a semantics not a semantics: some reasons for disliking the Routley–Meyer semantics for relevance logic," *Journal of Philosophical Logic* vol. 8, pp. 399–413.
Creswell, Max (1990) *Entities and Indices* (Dordrecht: Kluwer).
Davidson, Donald (1967) "Truth and meaning," *Synthese* vol. 17, pp. 304–23.
Davidson, Donald and Harman, Gilbert (1972) (eds.) *Semantics of Natural Language* (Dordrecht: Reidel).
Davidson, Donald and Hintikka, Jaakko (1969) (eds.) *Words and Objections: Essays on the Work of W. V. Quine* (Dordrecht: Reidel).
Demopoulos, William (1995) (ed.) *Frege's Philosophy of Mathematics* (Cambridge, MA: Harvard University Press).
Detlefsen, Michael (1992) (ed.) *Proof, Logic and Formalization* (London: Routledge).
Diogenes Laertius (1925) *Lives and Opinions of Eminent Philosophers*, translated from the Greek by R. D. Hicks, Loeb Classical Library (Cambridge, MA: Harvard University Press).
Dummett, Michael (1959) "Truth," in Dummett (1978), pp. 1–24.
(1973a) "The philosophical basis of intuitionistic logic," in Dummett (1978), pp. 215–47.
(1973b) "The justification of deduction," in Dummett (1978), pp. 290–318.
(1973c) "The significance of Quine's indeterminacy thesis," in Dummett (1978), pp. 375–419.
(1977) *Elements of Intuitionism* (Oxford: Oxford University Press).
(1978) *Truth and Other Enigmas* (Cambridge, MA: Harvard University Press).
Edgington, Dorothy (1985) "The paradox of knowability," *Mind* vol. 94, pp. 557–68.
Evans, Gareth and McDowell, John (1976) (eds.) *Truth and Meaning: Essays in Semantics* (Oxford: Oxford University Press).
Farber, M. (1950) (ed.) *Philosophic Thought in France and the United States* (Buffalo, NY: University of Buffalo Press).
Feferman, Solomon (1977) "Theories of finite type related to mathematical practice," in Barwise (1977), pp. 913–72.
Field, Hartry H. (1980) *Science Without Numbers: A Defense of Nominalism* (Princeton, NJ: Princeton University Press).
(1989) *Realism, Mathematics and Modality* (Oxford: Basil Blackwell).

Fine, A. and Lepin, J. (1990) *PSA 88* [Proceedings of the 1988 Convention of the Philosophy of Science Association], vol. II (East Lansing, MI: Philosophy of Science Association).

Fine, Kit (2002) *The Limits of Abstraction* (Oxford: Oxford University Press).

Fitch, Frederic (1949) "The problem of the morning star and the evening star," *Philosophy of Science* vol. 16, pp. 137–41.

(1950) "Attribute and class," in Farber (1950), pp. 640–7.

(1963) "A logical analysis of some value concepts," *Journal of Symbolic Logic* vol. 28, pp. 135–42.

Føllesdal, Dagfinn (1961) "Referential opacity and modal logic," Harvard University doctoral dissertation, reprinted as Føllesdal (1966).

(1965) "Quantification into causal contexts," in Cohen and Wartofsky (1965), pp. 263–74; reprinted in Linsky (1971a), pp. 52–62.

(1966) *Referential Opacity and Modal Logic, Filosofiske Problemer*, vol. XXXII (Oslo: Oslo Universitetsforlaget).

(1969) "Quine on modality," in Davidson and Hintikka (1969), pp. 175–85.

(1986) "Essentialism and reference," in Hahn and Schlipp (1986), pp. 97–113.

Forbes, Graeme (1995) review of Marcus (1993), *Notre Dame Journal of Formal Logic* vol. 36, pp. 336–9.

Frege, Gottlob (1879) *Begriffsschrift: eine der arithmetischen nachgebildete Formelsprache des reinen Denkens* (Halle: Louis Nebert).

(1884) *Die Grundlagen der Arithmetik: Eine logisch-mathematische Untersuchung über den Begriff der Zahl* (Breslau: Wilhelm Koebner).

(1893/1903) *Grundgesetze der Arithmetik, begriffsschriftlich abgeleitet*, 2 vols. (Jena: Pohle).

(1950) *The Foundations of Arithmetic*, translation of Frege (1884) from the German by J. L. Austin (London: Blackwell).

(1967) *Begriffsschrift*, translation from the German of Frege (1879) by S. Bauer-Mengelberg, in van Heijenoort (1967), pp. 1–82.

French, Peter A. and Wettstein, Howard K. (2001) (eds.) *Midwest Studies in Philosophy XXV: Figurative Language* (London: Blackwell).

Gabbay, D. and Guenthner, F. (1984) (eds.) *Handbook of Philosophical Logic*, vol. II: *Extensions of Classical Logic* (Dordrecht: Reidel).

Gabbay, D., Rahman, S., Symons, J., and van Bendegen, J. P. (2004) (eds.) *Logic, Epistemology, and the Unity of Science* (Dordrecht: Kluwer).

Garson, James (1984) "Quantification in modal logic," in Gabbay and Guenthner (1984), pp. 249–308.

Goodman, Nelson (1956) "A world of individuals," in Bochenski *et al.* (1956), pp. 197–210; reprinted in Benacerraf and Putnam (1964).

Goodman, Nelson and Quine, W. V. O. (1947) "Steps toward a constructive nominalism," *Journal of Symbolic Logic* vol. 12, pp. 97–122.

Goranko, Valentin (1994) "Refutation systems in modal logic," *Studia Logica* vol. 53, pp. 299–324.

Grandy, Richard (1982) review of Dummett (1973a), Prawitz (1977), etc. *Journal of Symbolic Logic* vol. 47, pp. 689–94.

Grice, H. P. and Strawson, P. F. (1956) "In defense of a dogma," *Philosophical Review* vol. 65, pp. 141–58.

Grzegorczyk, Andrzej (1967) "Some relational systems and the associated topological spaces," *Fundamenta Mathematicae* vol. 60, pp. 223–31.

Haack, Susan (1974) *Deviant Logic* (Cambridge: Cambridge University Press).

Hahn, L. E. and Schlipp, P. A. (1986) *The Philosophy of W. V. Quine* (LaSalle, IL: Open Court).

Hájek, Petr and Pudlak, Pavel (1998) *Metamathematics of First-Order Arithmetic* (Berlin: Springer).

Haldén, Søren (1963) "A pragmatic approach to modal theory," *Acta Philosophica Fennica* vol. 16, pp. 53–64.

Hale, Bob and Wright, Crispin (2001) *The Reason's Proper Study: Essays towards a Neo-Fregean Philosophy of Mathematics* (Oxford: Oxford University Press).

Hallett, Michael (1984) *Cantorian Set Theory and Limitation of Size* (Oxford: Clarendon Press).

Harman, Gilbert (1982) "Conceptual role semantics," *Notre Dame Journal of Formal Logic* vol. 23, pp. 242–56.

Heck, Richard G., Jr. (1996) "On the consistency of predicative fragments of Frege's *Grundgesetze der Arithmetik*," *History and Philosophy of Logic* vol. 17, pp. 209–20.

Heijenoort, Jean van (1967) (ed.) *From Frege to Gödel: A Sourcebook in Mathematical Logic, 1879–1931* (Cambridge, MA: Harvard University Press).

Hempel, Carl G. (1950) "Problems and changes in the empiricist criterion of meaning," *Revue Internationale de Philosophie* vol. 4, pp. 41–63.

Hersh, Reuben (1997) *What is Mathematics, Really?* (Oxford: Oxford University Press).

Hilbert, David (1925/1967) "On the infinite," translated from the German by Stefan Bauer-Mengelberg, in van Heijenoort (1967), pp. 367–92.

Hintikka, Jaakko (1963) "Modes of modality," *Acta Philosophica Fennica* vol. 16, pp. 65–82.

(1982) "Is alethic modal logic possible?" *Acta Philosophica Fennica* vol. 35, pp. 89–105.

Hintikka, Jaakko and Sandu, Gabriel (1995) "The fallacies of the new theory of reference," *Synthese* vol. 104, pp. 245–83.

Hofweber, Thomas and Everett, Anthony (2000) (eds.) *Empty Names, Fiction and the Puzzles of Non-Existence* (Chicago: CSLI).

Horsten, Leon and Halbach, Volker (2002) (eds.) *Principles of Truth* (Frankfurt: Hänsel-Hohenhausen).

Hrbacek, K. and Jech, T. (1999) *Introduction to Set Theory*, 3rd edn (New York: Marcel Dekker).

Hughes, G. E. and Creswell, M. J. (1968) *An Introduction to Modal Logic* (London: Methuen).

Hull, D., Forbes, M., and Okruhlik, K. (1993) (eds.) *PSA 92* [Proceedings of the 1992 Convention of the Philosophy of Science Association], vol. II (East Lansing, MI: Philosophy of Science Association).

Humphreys, P. and Fetzer, J. (1998) (eds.) *The New Theory of Reference*, Synthese Library vol. CCLXX (Dordrecht: Kluwer).
Irvine, Andrew (1989) (ed.) *Physicalism in Mathematics* (Dordrecht: Kluwer).
James, William (2000) "Pragmatism," in *Pragmatism and Other Writings*, ed. G. Gunn (New York: Penguin), pp. 1–132.
Jeffrey, Richard C. (1996) "Logicism 2000," in Stich and Morton (2002), pp. 1–132.
(2002) "Logicism lite," *Philosophy of Science* vol. 69, pp. 447–51.
Kahle, Reinhard (forthcoming) (ed.) *Intensionality: An Interdisciplinary Discussion* (Boston: A. K. Peters, Lecture Notes in Logic).
Katz, Jerrold (1985) (ed.) *The Philosophy of Linguistics* (Oxford: Oxford University Press).
Kitcher, Philip (1978) "The plight of the platonist," *Noûs*, vol. 12, pp. 119–36.
Klibansky, Raymond (1968) *Contemporary Philosophy*, 4 vols. (Florence: Editrice Nuova Italia).
Kneale, William and Kneale, Mary (1962) *The Development of Logic* (Oxford: Clarendon Press).
König, Julius (1905) "Über die Grundlagen der Mengenlehre und das Kontinuumproblem," *Mathematische Annalen* vol. 61, pp. 156–60.
(1967) "On the foundations of set theory and the continuum problem," translation of König (1905) from the German by Stefan Bauer-Mengelberg, in van Heijenoort (1967), pp. 145–9.
Kranz, D., Luce, R., Suppes, P., and Tversky, A. (1971) *Foundations of Measurement* (New York: Academic Press).
Kreisel, Georg (1967) "Informal rigour and completeness proofs," in Lakatos (1967), pp. 138–57.
Kripke, Saul (1963) "Semantical considerations on modal logic," *Acta Philosophica Fennica* vol. 16, pp. 83–94.
(1972) "Naming and necessity: Lectures give to the Princeton University Philosophy Colloquium, January, 1970," in Davidson and Harman (1972), pp. 253–355 and 763–9; reprinted with a new preface as Kripke (1980).
(1976) "Is there a problem about substitutional quantification?" in Evans and McDowell (1976), pp. 325–420.
(1979) "A puzzle about belief," in Margalit (1977), pp. 239–83.
(1980) *Naming and Necessity* (Cambridge, MA: Harvard University Press).
(1982) *Wittgenstein on Rules and Private Language* (Cambridge, MA: Harvard University Press).
Lavine, Shaughan (1995) review of Marcus (1993), *British Journal of the Philosophy of Science* vol. 46, pp. 267–74.
Lakatos, Imre (1967) (ed.) *Proceedings of the International Colloquium in the Philosophy of Science, London, 1965*, vol. I (Amsterdam: North Holland).
Lee, O. H. (1936) *Philosophical Essays for A. N. Whitehead* (New York: Longmans).
Lewis, David K. (1969) *Convention* (Cambridge, MA: Harvard University Press).
(1970) "Anselm and actuality," *Noûs* vol. 4, pp. 175–88.
(1991) *Parts of Classes* (Oxford: Oxford University Press).

Linsky, Leonard (1971a) (ed.) *Reference and Modality* (Oxford: Oxford University Press).

(1971b) "Essentialism, reference, and modality," in (Linsky (1971a), pp. 88–100.

(1977) *Names and Descriptions* (Chicago: University of Chicago Press).

Linström, Sten (forthcoming) (ed.) *Logicism, Intuitionism, Formalism. What Has Become of Them?* (Berlin: Springer).

Maddy, Penelope (1980) "Perception and mathematical intuition," *Philosophical Review* vol. 89, pp. 163–96.

(1984) "Mathematical epistemology: what is the question?" *The Monist* vol. 67, pp. 46–55.

(1990) "Mathematics and Oliver Twist," *Pacific Philosophical Quarterly* vol. 71, pp. 189–205.

Makinson, David (1966) "How meaningful are modal operators?" *Australasian Journal of Philosophy* vol. 44, pp. 331–7.

Manin, Yuri (1977) *A Course in Mathematical Logic*, translated from the Russian by N. Koblitz (Berlin: Springer).

Marcus, Ruth Barcan (1960) "Extensionality," *Mind* vol. 69, pp. 55–62.

(1963a) "Modalities and intensional languages," in Wartofsky (1963), pp. 77–96.

(1963b) "Attribute and class in extended modal systems," *Acta Philosophical Fennica* vol. 16, pp. 123–36.

(1967) "Essentialism in modal logic," *Noûs* vol. 1, pp. 90–6.

(1968) "Modal logic," in Klibansky (1968), pp. 87–101.

(1978) Review of Linsky (1977), *Philosophical Review* vol. 87, pp. 497–504.

(1990) "Some revisionary proposals about belief and believing," *Philosophy and Phenomenological Research* vol. 50 (Supplement), pp. 133–53.

(1993) *Modalities: Philosophical Essays* (Oxford: Oxford University Press).

Marcus, R. B., Quine, W. V., Kripke, S. A. *et al.* (1963) Discussion of Marcus (1963a) in Wartofsky (1963), pp. 105–16.

Margalit, Avishai (1977) *Meaning and Use* (Dordrecht: Reidel).

Martinich, A. P. (1979) *The Philosophy of Language* (Oxford: Oxford University Press).

Matiyasevich, Yuri (1993) *Hilbert's Tenth Problem* (Cambridge, MA: MIT Press).

McKinsey, J. C. C. (1941) "A solution to the decision problem for the Lewis systems S2 and S4, with an application to topology," *Journal of Symbolic Logic* vol. 6, pp. 117–34.

(1945) "On the syntactical construction of modal logic," *Journal of Symbolic Logic* vol. 10, pp. 83–96.

Metakides, George (1982) (ed.) *Proceedings of the First Patras Logic Symposion* (Amsterdam: North Holland).

Montagna, Franco and Mancini, Antonella (1994) "A minimal predicative set theory," *Notre Dame Journal of Formal Logic* vol. 35, pp. 186–203.

Müller, Gert-Heinz (1976) *Sets and Classes* (Amsterdam: North-Holland).

Nabokov, Vladimir (1980) *Lectures on Literature*, ed. F. Bowers (New York: Harcourt Brace Jovanovich).

Newman, J. R. (1956) (ed.) *The World of Mathematics*, 4 vols. (New York: Simon and Schuster).
Parsons, Charles (1980) "Mathematical intuition," *Proceedings of the Aristotelian Society*, vol. 80, pp. 145–68.
Parsons, Terence (1969) "Essentialism and quantified modal logic," *Philosophical Review* vol. 78, pp. 35–52; reprinted in Linsky (1971a), pp. 73–87.
(1987) "On the consistency of the first-order portion of Frege's logical system," *Notre Dame Journal of Formal Logic* vol. 28, pp. 61–8; reprinted in Demopoulos (1995), pp. 422–31.
Pollard, Stephen (1996) "Sets, wholes, and limited pluralities," *Philosophia Mathematica*, vol. 4, pp. 42–58.
Pour-El, M. and Richards, I. (1979) "A computable ordinary differential equation which possesses a computable solution," *Annals of Mathematical Logic* vol. 17, pp. 61–90.
(1981) "A wave equation with computable initial data such that its unique solution is not computable," *Advances in Mathematics* vol. 39, pp. 215–39.
(1983) "Noncomputability in analysis and physics: a complete determination of the class of noncomputable linear operators," *Advances in Mathematics* vol. 48, pp. 44–74.
(1987) "The eigenvalues of an effectively determined self-adjoint operator are computable, but the sequence of eigenvalues is not," *Advances in Mathematics* vol. 63, pp. 1–41.
Prawitz, Dag (1977) "Meaning and proofs: on the conflict between classical and intuitionistic logic," *Theoria* vol. 43, pp. 2–40.
Prior, Arthur N. (1960) "The runabout inference-ticket," *Analysis* vol. 21, pp. 38–9.
(1963) "Is the concept of referential opacity really necessary?" *Acta Philosophical Fennica* vol. 16, pp. 189–99.
(1967a) "Logic, modal," in Weiss (1967), vol. V, pp. 5–12.
(1967b) *Past, Present, and Future* (Oxford: Clarendon Press).
Putnam, Hilary (1971) *Philosophy of Logic* (New York: Harper).
(1975) "Truth and necessity in mathematics," in *Mathematics, Matter, and Method* (Cambridge: Cambridge University Press), pp. 1–11.
Quine, W. V. O. (1936) "Truth by convention," in Lee (1936), pp. 90–124.
(1946) Review of Barcan (1946), *Journal of Symbolic Logic* vol. 11, pp. 96–7.
(1947a) "The problem of interpreting modal logic," *Journal of Symbolic Logic* vol. 12, pp. 43–8.
(1947b) Review of Barcan (1947), *Journal of Symbolic Logic* vol. 12, pp. 95–6.
(1951a) "Carnap's views on ontology," *Philosophical Studies* vol. 2, pp. 65–72.
(1951b) "Two dogmas of empiricism," *Philosophical Review* vol. 60, pp. 20–43.
(1953) *From a Logical Point of View* (Cambridge, MA: Harvard University Press).
(1960) *Word and Object* (New York: John Wiley and Sons).
(1961) *From a Logical Point of View*, 2nd edn (Cambridge, MA: Harvard University Press).
(1963) Comments on Marcus (1963a) in Wartofsky (1963), pp. 97–104.

(1969) "Reply to Sellars," in Davidson and Hintikka (1969), p. 338.
(1970) *Philosophy of Logic* (Englewood Cliffs, NJ: Prentice-Hall).
(1980) *From a Logical Point of View*, 3rd edn (Cambridge, MA: Harvard University Press).
(1981) "Response to David Armstrong," in *Theories and Things* (Cambridge: Harvard University Press), pp. 182–4.
Ramsey, Frank Plumpton (1925) "The foundations of mathematics," *Proceedings of the London Mathematical Society* vol. 25, pp. 338–84.
Rayo, Augustin and Uzquiano, Gabriel (1999) "Towards a theory of second-order consequence," *Notre Dame Journal of Formal Logic* vol. 40, pp. 315–25.
Rescher, Nicholas (1968) (ed.) *Studies in Logical Theory* (Oxford: Basil Blackwell).
Richman, F. (1981) (ed.) *Constructive Mathematics* [Springer Lecture Notes in Mathematics 873] (Berlin: Springer).
(1975) *Logic Colloquium '73* (Amsterdam: North Holland).
Rosen, Gideon and Burgess, John P. (2005) "Nominalism reconsidered," in Shapiro (2005), pp. 460–82.
Rükert, Helge (2004) "A solution to Fitch's paradox of knowability," in Gabbay *et al.* (2004), pp. 351–80.
Russell, Bertrand (1902/1967) letter to Frege, translated from the German by Beverly Woodward, in van Heijenoort (1967), pp. 124–5.
(1985) *The Philosophy of Logical Atomism*, ed. David Pears (LaSalle, IL: Open Court).
Salerno, J. (2008) *New Essays on the Knowability Paradox* (Oxford: Oxford University Press).
Salmon, Nathan (1986) *Frege's Puzzle* (Cambridge, MA: MIT Press).
Schindler, Ralf-Dieter (1994) "A dilemma in the philosophy of set theory", *Notre Dame Journal of Formal Logic* vol. 35, pp. 458–63.
Schirn, Matthias (1998) (ed.) *Philosophy of Mathematics Today* (Oxford: Oxford University Press).
Scroggs, Schiller Joe (1951) "Extensions of the Lewis system S5," *Journal of Symbolic Logic* vol. 16, pp. 112–20.
Searle, John R. (1967) "Proper names and descriptions," in Weiss (1967) vol. VI, pp. 487–91.
(1979) "Metaphor," in Martinich (1979), pp. 92–123.
Shahan, R. W. and Swoyer, C. (1979) (eds.) *Essays on the Philosophy of W. V. Quine* (Norman: University of Oklahoma Press).
Shapiro, Stewart (1985) (ed.) *Intensional Mathematics* (Amsterdam: North Holland).
(1987) "Principles of reflection and second-order logic," *Journal of Philosophical Logic* vol. 16, pp. 309–33.
(1997) *Philosophy of Mathematics: Structure, Ontology, Modality* (Oxford: Oxford University Press).
(2005) (ed.) *The Oxford Handbook of Philosophy of Mathematics and Logic* (Oxford: Oxford University Press).
Skinner, B. F. (1957) *Verbal Behavior* (New York: Appleton-Century-Crofts).

Skura, Tomasz (1995) "A Lukasiewicz-style refutation system for the modal logic S4," *Journal of Philosophical Logic* vol. 24, pp. 573–82.

Slupecki, Jerzy and Bryll, Grzegorz (1973) "Proof of the L-decidability of Lewis system S5," *Studia Logica* vol. 24, pp. 99–105.

Smullyan, Arthur (1947) review of Quine (1947a), *Journal of Symbolic Logic* vol. 12, pp. 139–41.

(1948) "Modality and description," *Journal of Symbolic Logic* vol. 13, pp. 31–7.

Soames, Scott (1984) "What is a theory of truth?" *Journal of Philosophy* vol. 84, pp. 411–29.

(1985) "Semantics and psychology," in Katz (1985), pp. 204–26.

Stalnaker, Robert (1968) "A theory of conditionals," in Rescher (1968), pp. 98–112.

Stanley, Jason (2001) "Hermeneutic fictionalism," in French and Wettstein (2001), pp. 36–71.

Stich, S. and Morton, A. (2002) (eds.) *Benacerraf and his Critics* (London: Blackwell).

Stigt, W. P. van (1979) "The rejected parts of Brouwer's dissertation on the foundations of mathematics," *Historia Mathematica* vol. 6, pp. 385–404.

Tarski, Alfred (1931) "Sur les ensembles définissables de nombres réels," *Fundamenta Mathematicae* vol. 17, pp. 210–39.

(1935) "Der Wahrheitsbegriff in den formalisierten Sprachen," *Studia Philosophica* vol. 1, pp. 261–405.

(1936) "Über den Begriff der logischen Folgerung," in *Actes du Congrès International de Philosophie Scientifique*, vol. VII (Paris: Hermann), pp. 1–11.

(1944) "The semantic conception of truth," *Philosophy and Phenomenological Research* vol. 4, pp. 341–75.

(1983a) *Logic, Semantics, Metamathematics*, 2nd edn, ed. J. Corcoran (Indianapolis: Hackett).

(1983b) "On definable sets of real numbers," translation of Tarski (1931) from the French by J. H. Woodger, in Tarski (1983a), pp. 110–42.

(1983c) "The concept of truth in formalized languages," translation of Tarski (1935) from the German by J. H. Woodger, in Tarski (1983a), pp. 152–278.

(1983d) "On the concept of logical consequence," translation of Tarski (1936) from the German by J. H. Woodger, in Tarski (1983a), pp. 409–20.

Tarski, Alfred and Vaught, Robert (1956) "Arithmetical extensions of relational systems," *Compositio Mathematica* vol. 13, pp. 81–102.

Tarski, A., Mostowski, A., and Robinson, R. M. (1953) *Undecidable Theories* (Amsterdam: North Holland).

Thomas, Robert (2000) "Mathematics and fiction I: identification," *Logique et Analyse* vol. 43, pp. 301–40.

(2002) "Mathematics and fiction II: analogy," *Logique et Analyse* vol. 45, pp. 185–228.

Thomason, R. H. (1984) "Combination of tense and modality," in Gabbay and Guenthner (1984), pp. 135–65.

Thomason, S. K. (1973) "A new representation of S5," *Notre Dame Journal of Formal Logic* vol. 14, pp. 281–7.

Thomson, J. J. (1987) *On Being and Saying: Essays for Richard Cartwright* (Cambridge, MA: MIT Press).

Tomberlin, James E. (1994) (ed.) *Logic and Language* [*Philosophical Perspectives*, vol. VIII] (Atascadero, CA: Ridgeview Publishing).

Uzquiano, Gabriel (2003) "Plural quantification and classes," *Philosophia Mathematica*, vol. 11, pp. 67–81.

Wartofsky, Max (1963) (ed.) *Proceedings of the Boston Colloquium for the Philosophy of Science 1961/1962* (Dordrecht: Reidel).

Wehmeier, Kai (forthcoming) "Modality, mood, and descriptions," to appear in Kahle (forthcoming).

Weiss, Paul (1967) (ed.) *Encyclopedia of Philosophy*, 6 vols. (New York: Macmillan).

Weyl, Hermann (1944) "David Hilbert and his Mathematical Work," *Bulletin of the American Mathematical Society* vol. 50, pp. 612–54.

White, Leslie A. (1947) "The locus of mathematical reality: an anthropological footnote," *Philosophy of Science* vol. 14, pp. 289–303, reprinted in Newman (1956), vol. IV.

Whorf, Benjamin Lee (1956) "Science and linguistics," in *Language, Thought, and Reality: Selected Writings*, ed. J. B. Carroll (Cambridge, MA: MIT Press), pp. 207–19.

Williamson, Timothy (1987) "On the paradox of knowability," *Mind* vol. 96, pp. 256–61.

Woods, John (2002) *Paradox and Paraconsistency: Conflict Resolution in the Abstract Sciences* (Cambridge: Cambridge University Press).

Wright, Crispin (1980) *Wittgenstein on the Foundations of Mathematics* (Cambridge: Cambridge University Press).

(1983) *Frege's Conception of Numbers as Objects* (Aberdeen: Scots Philosophical Monographs).

Wright, G. H. von (1951) *An Essay in Modal Logic* (Amsterdam: North Holland).

Yablo, Steven (2000) "A paradox of existence," in Hofweber and Everett (2000), pp. 275–311.

Index